Meshfree and Particle Based Approaches
in Computational Mechanics

Meshfree & Particle Based Approaches in Computational Mechanics

edited by
Piotr Breitkopf & Antonio Huerta

Publisher's note

Every possible effort has been made to ensure that the information contained in this book is accurate at the time of going to press, and the publishers and authors cannot accept responsibility for any errors or omissions, however caused. No responsibility for loss or damage occasioned to any person acting, or refraining from action, as a result of the material in this publication can be accepted by the editor, the publisher or any of the authors.

First published in Great Britain and the United States in 2004 by Kogan Page Limited

First South Asian Edition 2007

Kogan Page Limited
120 Pentonville Road
London N1 9JN
United Kingdom
www.kogan-page.co.uk

Kogan Page India
4737/23 Ansari Road
New Delhi- 110002

© Hermes Science and Kogan Page Limited

ISBN 1-9039-9645-7

British Library Cataloguing-in-Publication Data

A CIP record for this book is available from the British Library.

Library of Congress Cataloging-in-Publication Data

Meshfree and particle based approaches in computational mechanics/edited by Piotr Breitkopf & Antonio Huerta.
   p. cm.
Includes index.
   ISBN 1-9039-9645-7
   1. Mechanics, Analytic - Data processing. 2. Meshfree methods (Numerical analysis) I. Breitkopf, Piotr. II. Huerta, Antonio, 1959-
QA808.M45 2004
531--dc22
                                      2004007789

Typeset by Newgen Imaging Systems (P) Ltd., Chennai, India
Printed in Brijbasi Art Press Ltd., I-72, Sector-9, Noida, U.P. India.

Contents

Foreword

This publication is dedicated to meshfree and particle-based methods in computational mechanics.

Is there a unique meshfree technique? In recent years, we have witnessed the development of a large family of methods whose objective is to get rid of mesh constraints. Even if no definite answers are yet established, the results spur further research. In fact, this area of research is evolving quickly towards what may become in future years a new generation of computational methods in engineering and applied sciences. This comparatively short publication cannot pretend to give a comprehensive view of a subject whose present success is precisely due to the abundance of theoretical approaches and applications. Instead of an exhaustive presentation we have chosen to cover different approaches and sensibilities. In fact, we have voluntarily limited ourselves to several European contributions. Meshfree techniques are still an emerging topic. The papers focus on fundamental ideas, which are illustrated by a variety of applications. You will find formulations based on moving least squares, smooth particle hydrodynamics and generalized finite differences combined with applications in acoustics, fluid and solid mechanics, as well as numerical and experimental data smoothing.

Meshfree methods were initiated in the early 1970s by the group led by Professor Janusz Orkisz at the Cracow University of Technology.

A paper from the University of Technology of Compiègne in 1992 revisited the field by introducing the concept of diffuse elements. Ten years later, the early research groups are still active in the field and we find their contributions included here. The papers from six European countries: Poland, Belgium, Germany, Great Britain, France and Spain testify the vitality of this research area in Europe.

Piotr Breitkopf
Université de Technologie de Compiègne

Antonio Huerta
Universidad Politécnica de Cataluña, Barcelona

Chapter 1

An Introduction to Moving Least Squares Meshfree Methods

Piotr Breitkopf, Alain Rassineux & Pierre Villon
Laboratoire de Mécanique Roberval
Université de Technologie de Compiègne, France

1. Introduction

Meshfree techniques provide a promising alternative to solving partial differential equations (PDE) using finite elements. The main feature of meshfree methods is the absence of an explicit mesh. *Smooth particle hydrodynamics* (SPH, Lucy, 1977) can be recognized as one of the first meshfree approaches. We do not pretend here to provide an exhaustive survey of the domain; there are comprehensive works (Belytchko *et al.*, 1996 and Babuska *et al.*, 2002) that may be consulted for reference. Here, we provide a quick historical review.

Two main families of methods can be distinguished. The first group involves collocation methods: *the generalized finite difference method* (*GFDM, Liszka and Orkisz, 1980), particle in cell* (Sulsky and Schreyer, 1993), the *finite point method* (Onate and Idesohn, 1998), the *double grid collocation* (Breitkopf *et al.*, 2000) and the *least squares collocation method* (Zhang *et al.*, 2001). Galerkin-like methods were introduced through the *element method*, (DEM, Nayroles *et al.*, 1992) followed by the *element-free Galerkin method* (EFG, Belytschko *et al.*, 1994) and the *reproducing kernel particle method* (RKPM, Liu *et al.*, 1996). More recently variational SPH (Bonet and Lok, 1999), the *meshless local Petrov-Galerkin* (MLPG, Lin and Atluri, 2000), and the *method of finite spheres* (De and Bathe, 2000) appeared. The methods of *extended finite element method* (XFEM, Sukumar *et al.*, 2000) and the *partition of unity method* (PUFEM, Babuska and Melenk, 1997) are not presented here.

In this paper, we focus on meshfree methods using *moving least squares* (MLS) techniques. The origins of MLS approximation can be found in independent works in several fields. In the domain of geostatistics, we find the early concept of weighted moving approximation in the work of Krige which gave rise to the term of *kriging* introduced later by Matheron (Krige, 1966, Matheron, 1963). In the field of non-parametric estimation in statistics, this work (Cleveland, 1979) is based on similar principles. In the field of smoothing data, we note the development of methods of approximation without solving a global system (Shepard, 1968, Mac Lain, 1974, Barnhill, 1977, Gordon *et al.*, 1978). The term 'moving least squares' was introduced by (Lancaster and Salkauskas, 1981).

The fundamental idea behind MLS meshfree concepts aims at a better control of shape function smoothness and continuity as opposed to finite elements. This is obtained through the use of the weight functions. The weight functions are associated with a node and their values decrease with the distance. They allow control of the locality and the continuity of the approximation. The MLS equivalent of the shape functions is derived from a minimization of a weighted least squares criterion. The difference between weights and shape functions is that the shape functions satisfy the consistency conditions necessary for the numerical solution of the PDEs. We call this process the "shape functions factory".

In the Galerkin approach, several problems have to be solved. First, the weak form requires numerical integrations on the boundary and inside the domain. Several authors propose different strategies. The "truly meshless" techniques (Lin and Atluri, 2000, De and Bathe, 2000) can be contrasted with domain decomposition techniques (Nayroles *et al.*, 1992, Belytschko *et al.*, 1994). An intermediary method is based on nodal integration (Beissel and Belytschko, 1996, Bonet and Lok, 1999, Chen *et al.*, 2002). The essential boundary conditions can be taken into account using nodal interpolation (Nayroles *et al.*, 1992), using Lagrange multipliers (Belytschko *et al.*, 1994) or several modified variational principles (Babuska *et al.*, 2002).

This paper is organized as follows. Section 2 details different formulations and implementation issues involved in obtaining a robust MLS approximation. The theoretical and numerical convergence is considered. In section 3 we describe the shape function factory. We also give the explicit form of the shape functions in 1D, 2D and 3D. In section 4, we establish the interpolation property for a general case and we give implementation details. The next section discusses different strategies for the choice of domains of influence. Section 6 is devoted to the integration scheme with respect to integration constraints.

2. Moving least squares approximation

We first introduce the *moving least squares* (MLS) approximation following the approach (Lancaster and Salkauskas, 1986) which may be interpreted (Nayroles *et al.*, 1992) as a generalization of the finite elements. An alternative method (Liszka and Orkisz, 1980), based on a local Taylor expansion reveals numerous advantages and can also be used.

We look for a local approximation of a function u_{ex} at a point x, based on the nodal values u_i of the function u_{ex} at a limited number of points x_i close to x. The unknown function u_{ex} is approximated in the vicinity of x by

$$u_{ex}(x) \approx u_{app}(x) = \mathbf{p}^T(x)\mathbf{a}(x) \tag{1}$$

The most often-used are polynomial basis functions

$$\mathbf{p}^T(x) = [1 \ x \ \cdots \ x^n] \tag{2}$$

although the use of other functions, for instance trigonometric functions, has also been investigated (Belytschko *et al.*, 1994a, Savignat, 2000).

Coefficients a_i of the approximation are related to the nodal values u_i by minimizing a norm of the weighted difference between the estimated values at nodes and the nodal values u_i.

$$J_x(\mathbf{a}) = \frac{1}{2} \sum_i w_i(x_i, x) \left(\mathbf{p}^T(x_i)\mathbf{a} - u_i \right)^2 \tag{3}$$

The contribution of each nodal value to the approximation is influenced by a weighting function $w(x_i, x)$ such that $w(x_i, \cdot) > 0$ inside the domain of influence of the node i and $w(x_i, \cdot) = 0$ otherwise, providing a local character to the approximation. We discuss the issues relative to the construction and to the choice of different weighting functions in section 2.2 of this paper.

Generally, MLS formulation does not interpolate data, therefore the relation

$$u_{app}(x_i) = u_{ex}(x_i) \tag{4}$$

is not verified. The interpolation property (4) is commonly obtained with the weighting functions which take infinite value at the node

$$x \to x_i \ \Rightarrow \ w(x_i, x) \to \infty \tag{5}$$

In this case the influence of other nodes vanishes, the approximation becomes interpolating and (4) is satisfied. In section 4 of this paper, we discuss thoroughly this issue and we present a method in order to obtain non-singular interpolating weight functions. Another way of enforcing interpolation in the context of RKPM was recently proposed (Chen *et al.*, 2002). We remark that, contrary to the finite element interpolation, the MLS interpolation property is not sufficient for the enforcement of the essential boundary conditions. In a finite element context, the influence of the internal nodes vanishes at the boundary and the interpolation depends only on the boundary nodal values. Because of the construction of the MLS approximation itself, this property is not preserved. Thus, special treatment is needed and one of the techniques used to enforce essential boundary conditions is illustrated in section 6.

2.1. The "diffuse" and "full" derivatives

The derivatives of $u_{app}(x)$ may be approximated in two ways. The first form is a "full derivative", denoted by du_{app}/dx and is obtained by standard derivation of both $\mathbf{p}(x)$ and $\mathbf{a}(x)$ in (1):

$$\frac{du_{app}}{dx} = \frac{d\mathbf{p}^T}{dx}(x)\mathbf{a}(x) + \mathbf{p}^T(x)\frac{d\mathbf{a}}{dx} \tag{6}$$

The second form is obtained by considering that coefficients $\mathbf{a}$ are constant; this leads to the "diffuse derivative" denoted by $\frac{\delta}{\delta x}$

$$\frac{\delta u_{app}}{\delta x}(x) = \frac{d\mathbf{p}^T}{dx}(x)\mathbf{a}(x) \tag{7}$$

The former approach is used in the element-free Galerkin method (Belytschko et al., 1994) and the latter one is analogous to the derivatives obtained by the GFDM method (Liszka and Orkisz 1980) where second order diffuse derivatives are employed. The first order diffuse derivative was reintroduced by (Nayroles et al., 1992) along with the diffuse element method. Both derivatives converge to the exact ones when the discretization size tends to zero.

The two derivatives are equivalent in the three following cases:

- evaluation point x is located at a node and the interpolating condition (4) is verified (see section 4),
- weights $w(x_i, x)$ are constant over a vicinity of x: in this case, coefficients $\mathbf{a}$ are constant, the term $d\mathbf{a}/dx$ vanishes and $(\delta u_{app}/\delta x)(x) \equiv (du_{app}/dx)(x)$,
- u may be expressed as a linear combination of basis functions p_i: coefficients $\mathbf{a}$ are constant in this case too.

The function converges to the first terms of its Taylor expansion when the discretization size tends to zero. Therefore, for an arbitrary function, the equivalence between the two derivatives is obtained in the limit.

The "diffuse derivative" may be intuitively interpreted as an approximation to the derivative of the function u_{ex}, while the "full derivative" is the derivative of the approximated function u_{app}. Both types of derivatives present drawbacks and advantages and the choice depends on the application. In section 2.5 we develop the interpretation of the diffuse derivative in the terms of Taylor series expansion of function $u(x)$ and we demonstrate convergence properties.

2.2. Weighting functions

Fundamental properties related to MLS approximation, such as locality and continuity, mainly depend on an appropriate choice of the weighting functions w_i. In order to limit the number of nodes used for the local evaluation, the support of the approximation must be bounded. As a consequence, the bandwidth of the resulting global linear system is also reduced. The weight function vanishes at a finite distance from x_i, called *radius of influence* and denoted as r (for details see below section 2.2.3). The area around x_i is called *domain of influence* of node x_i. Function w_i has a maximum (usually unit) value at node x_i, remains positive and decreases continuously over the domain of influence. The choice among weight functions satisfying the above requirements depends on the application at hand. In particular, this choice is influenced by the required degree of continuity of the approximation. Whenever a "full" derivative approach is used, differentiable weights must be chosen.

2.2.1. Window functions

The weight functions are constructed from the reference window functions w_{ref}. When using the diffuse derivative, we do not differentiate the weights and the choice of the basic hat function

$$w_{ref} = \begin{cases} 1 - s, & s < 1 \\ 0, & s \geq 1 \end{cases} \tag{8}$$

is straightforward. The choice of w_{ref} is also driven by performance purposes. For this reason, spline functions are preferred rather than exponentials or trigonometric functions. A C^1 piecewise cubic spline coefficient may for instance be computed to ensure the following conditions

$$w_{ref}(0) = 1, \quad w_{ref,x}(0) = w_{ref}(1) = w_{ref,x}(1) = 0$$

giving the expression

$$w_{ref}(s) = \begin{cases} 1 - 3s^2 + 2s^3, & 0 \leq s < 1 \\ 0, & s \geq 1 \end{cases} \tag{9}$$

Figure 1 and Figure 2 show the examples of 1D reference window functions given respectively by the formulae (8) and (9).

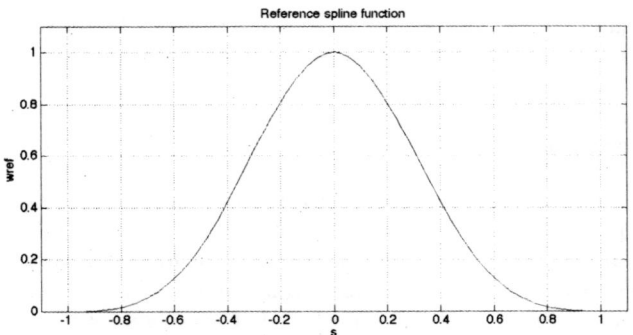

Figure 1 *Hat reference window function*

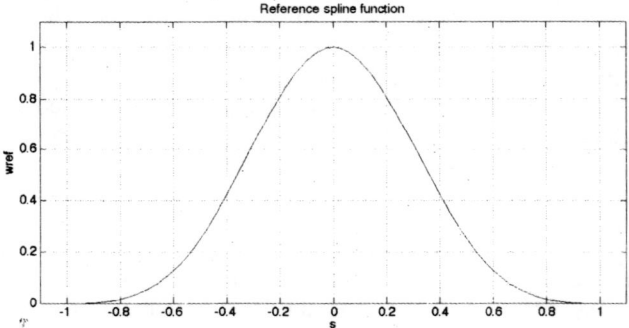

Figure 2 *Spline reference window function*

2.2.2. Weight functions

Weight functions are obtained from the reference window functions by substituting the relative distance between the evaluation point and the node.

$$w(x) = w_{ref}\left(\frac{dist(x_i, x)}{r}\right) \tag{10}$$

2D and 3D weight functions can be obtained directly from w_{ref} by the use of an appropriate norm to compute the relative distance $dist(x_i, x)/r$. Another way consists in using a tensor product of one-dimensional weights

$$w_{2D}(x, y) = w_{ref}\left(\frac{|x_i - x|}{r_x}\right) w_{ref}\left(\frac{|y_i - y|}{r_y}\right)$$

$$w_{3D}(x, y, z) = w_{ref}\left(\frac{|x_i - x|}{r_x}\right) w_{ref}\left(\frac{|y_i - y|}{r_y}\right) w_{ref}\left(\frac{|z_i - z|}{r_z}\right) \tag{11}$$

2.2.3. Domains of influence

We call "domain of influence" of node i, the adherence of the set $\Omega_i = \{x/w(x_i, x) > 0\}$. Two different strategies are possible for establishing the "radius of influence" r appearing in the equation (10):

–   at each evaluation point we take into account k closest nodes – this method is referred to as the $R(x)$ strategy;
–   the domains of influence are arbitrarily fixed by assigning a radius of influence to each node – this method is referred to as the r_i strategy.

Figure 3, Figure 4 and Figure 5 show different forms of domains of influence of the central node using various definitions of the radius of influence r. In all three cases $n_\mathbf{p} = 3$ and the radius of influence is chosen in such a way that at least 4 closest neighbors are selected. $n_\mathbf{p}$ is the number of terms of the polynomial basis vector $\mathbf{p}$. A regular 2D grid is used in first two figures. The Figure 3 represents $R(x)$ strategy. In this case, the domain of influence is the union of 4th order Voronoi cells connected with the central node. Figure 4 shows the r_i strategy combined with L^2 norm which results in a circular domain. A randomly perturbed grid is used in Figure 5, where the L^∞ norm is employed in order to get a square domain of influence.

The existence of the approximation requires a number of nodes at least equal to $n_\mathbf{p}$ at each evaluation point. When $n = n_\mathbf{p}$, MLS degenerates to polynomial Lagrange interpolation and the weights have effect any longer. So, in order to guarantee the continuity, the size of the domains of influence must be adjusted. In a general case, at least $n_\mathbf{p} + $ dim nodes are recommended at each point of the domain, where "dim" is the space dimension.

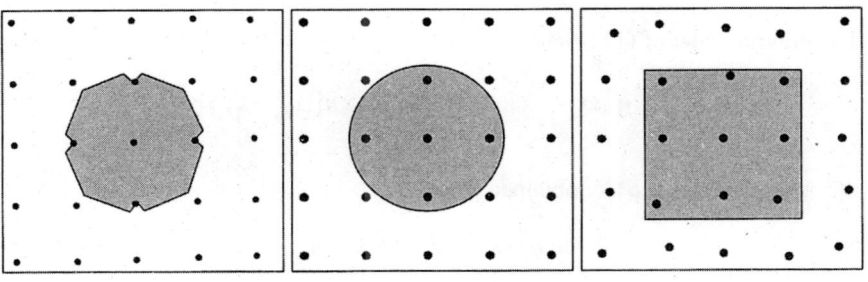

Figure 3	**Figure 4**	**Figure 5**
$R(x)$ strategy	*r_i strategy, L^2 norm*	*r_i strategy, L^∞ norm*

Different forms of domains of influence of the central node using the various definitions of the radius of influence r on a regular 2D (Figure 3, Figure 4) grid and on a randomly perturbed grid (Figure 5) with $n_\mathbf{p} = 3$ and at least 4 closest neighbors

2.3. *Centered moving least squares*

Let us introduce a polynomial basis $\mathbf{q}$ centered at the evaluation point x. For a node x_i we have

$$\mathbf{q}^T(x_i - x) = \begin{bmatrix} 1 & (x_i - x) & \cdots & \dfrac{(x_i - x)^k}{k!} \end{bmatrix} \tag{12}$$

For $k = 2$, the new basis $\mathbf{q}$ is related to the basis $\mathbf{p}$ by the following relationship

$$\mathbf{Q}\mathbf{p}(x_i) = \mathbf{q}(x_i - x), \quad \mathbf{Q} = \begin{vmatrix} 1 & 0 & 0 \\ -x & 1 & 0 \\ \frac{1}{2}x^2 & -x & \frac{1}{2} \end{vmatrix} \tag{13}$$

and inversely, the basis $\mathbf{p}$ is related to the basis $\mathbf{q}$, centered at x_i by

$$\mathbf{p}(x_i) = \mathbf{Q}^{-1}\mathbf{q}(x_i - x), \quad \mathbf{Q}^{-1} = \begin{vmatrix} 1 & 0 & 0 \\ x & 1 & 0 \\ x^2 & 2x & 2 \end{vmatrix} \tag{14}$$

The matrix $\mathbf{Q}$ is nonsingular as it corresponds to a basis change in a polynomial vector space. The nodal approximation (1) becomes

$$u_{app}(x_i) = \mathbf{q}^T(x_i - x)\mathbf{Q}^{-T}(x)\mathbf{a}(x) = \mathbf{q}^T(x_i - x)\alpha(x) \tag{15}$$

where

$$\alpha(x) = \mathbf{Q}^{-T}(x)\mathbf{a}(x) \tag{16}$$

The insertion (16) into criterion (3) leads to a modified criterion which depends on vector α

$$J_x(\alpha) = \frac{1}{2}\sum_j w(x_j, x)(\mathbf{q}^T(x_j - x)\alpha - u_j)^2 \tag{17}$$

The minimization of (17) yields

$$\sum_j (\mathbf{q}^T(x_j - x)\alpha - u_j)w(x_j, x)\mathbf{q}(x_j - x) = \mathbf{0} \tag{18}$$

thus the coefficients α are obtained from

$$\alpha(x) = \mathbf{A}(x)^{-1}\mathbf{B}(x)\mathbf{u} \tag{19}$$

where $\mathbf{A}$ and $\mathbf{B}$ matrices are given by the following formulae:

$$\mathbf{A}(x) = \sum_i w_i \mathbf{q}(x_i - x)\mathbf{q}^T(x_i - x)$$

$$\mathbf{B}(x) = [\cdots \quad w_i \mathbf{q}(x_i - x) \quad \cdots] \tag{20}$$

The algorithms based on the centered approach exhibit better conditioning properties then those using the global coordinates. In the centered approach, the condition number of the matrix $\mathbf{A}$ does not depend on the absolute position of the set of nodes.

By computing the consecutive terms of the matrix-vector product (16), we find that coefficients α are the diffuse derivatives of the approximation as introduced in (7)

$$
\begin{aligned}
\alpha_0 &= a_0 + a_1 x + a_2 x^2 = \mathbf{pa} = u_{app}(x) \\
\alpha_1 &= a_1 + 2a_2 x = \frac{d\mathbf{p}}{dx}\mathbf{a} = \frac{\delta u_{app}}{\delta x}(x) \\
\alpha_2 &= 2a_2 = \frac{d^2\mathbf{p}}{dx^2}\mathbf{a} = \frac{\delta^2 u_{app}}{\delta x^2}(x)
\end{aligned}
\tag{21}
$$

2.4. Dimensionless moving least squares

The centered approach gives better conditioned matrices $\mathbf{A}$ than formulations expressed in a global coordinate system. However, when the characteristic size of the nodal pattern decreases, near-singular matrices are obtained. The conditioning can be further improved by introducing local dimensionless coordinates $\xi_i = \frac{x_i - x}{h}$. Scaling factor h is chosen in such a way that $0 \le \xi \le 1$, for instance $h = \max(dist(\mathbf{x}_i, \mathbf{x}))$. We write

$$
\mathbf{D}(h)\mathbf{p}(x_i - x) = \mathbf{p}(\xi_i)
\tag{22}
$$

where

$$
\mathbf{D} =
\begin{bmatrix}
1 & & & 0 \\
& 1/h & & \\
& & \ddots & \\
0 & & & 1/h^k
\end{bmatrix}
\tag{23}
$$

Cost function (17) is expressed in the dimensionless coordinate system as

$$
J_x(\beta) = \frac{1}{2}\sum_i w(x_i, x)(\mathbf{q}^T(\xi)\beta - u_i)^2
\tag{24}
$$

The relationship between α and β coefficients is given by diagonal matrix $\mathbf{D}$ introduced in (23)

$$
\alpha = \mathbf{D}(h)\beta
\tag{25}
$$

We show in section 2.6 why this formulation should be preferred in practical programming.

2.5. *Convergence of the MLS approximation*

In this section we show that the diffuse derivatives (7) correspond to an approximation of a Taylor series expansion. Let us consider that function $u_{ex}(\cdot)$ is $k+1$ times continuously differentiable. It can be proved (Villon 1991) that the vector of coefficients α converges to the vector of the "full derivatives" (6) generalized for an arbitrary order of derivation k

$$\mathbf{U}_{ex}(x) = \left[u_{ex}(x), \frac{du_{ex}}{dx}(x), \ldots, \frac{d^k u_{ex}}{dx^k}(x) \right]^T \qquad (26)$$

The Taylor expansion of $u_{ex}(x)$ in the vicinity of point x gives

$$u(x_i) = \mathbf{q}^T (x_i - x) \mathbf{U}_{ex}(x) + \varepsilon_i \qquad (27)$$

where ε is the truncation error. We substitute (27) into (17) and we get a criterion

$$\mathbf{J}_x(\alpha) = \frac{1}{2} \sum_j w(x_j, x)(\mathbf{q}^T (x_j - x)(\mathbf{U}_{ex} - \alpha) + \varepsilon_j)^2 \qquad (28)$$

We perform the minimization of $J_x(\alpha)$ and we introduce the dimensionless coordinates (22), the associated polynomial basis (23) and the matrix $\mathbf{A}$ (20) which gives

$$\mathbf{A}(\xi)\mathbf{D}^{-1}(\mathbf{U}_{ex} - \alpha) = \sum_i w_i \varepsilon_i \mathbf{q}(\xi_i) \qquad (29)$$

We note, that for a given nodal pattern, the dimensionless coordinates ξ do not depend on the pattern size r, thus $\mathbf{A}(\xi)$ and $\mathbf{q}(\xi_i)$ are constant too. Using the interpretation (21) of the subsequent coefficients α, the equation (26) and the fact that $\varepsilon_i = \xi_i^{k+1} \frac{u_{ex}^{(k+1)}(x, x_i)}{(k+1)!}$, we conclude that the error of the l'th "diffuse derivative" is bounded by

$$\left| \alpha_l(x) - \frac{d^l u_{ex}}{dx^l} \right| < \frac{r(x)^{k-l+1}}{(l+1)!} \left\| \frac{d^{k+1} u_{ex}}{dx^{k+1}} \right\| K(x) \qquad (30)$$

In the above formula, k is the degree of the polynomial basis, r is the characteristic size of the nodal pattern. The term $\| d^{k+1} u_{ex}/dx^{k+1} \|$ depends on the regularity of the function $u_{ex}(x)$ and the term $K(x)$ is related to the local topological "quality" of the pattern. When a fixed pattern of points is used and the radius r decreases, the order of convergence of the approximation of the k-th derivative is $0(k-l)$. The value of the term $K(x)$ is related to the conditioning of the system $\mathbf{A}(\xi)^{-1}\mathbf{B}(\xi)$ and depends on the local nodal pattern taken into account in the approximation at point x. The following figures illustrate different cases:

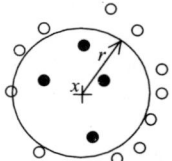

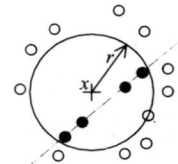

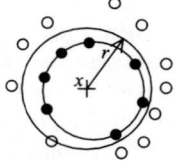

Figure 6 *Well conditioned pattern with linear basis*
$$\mathbf{p} = [1 \quad x \quad y]$$

Figure 7 *Pathological pattern with linear basis*
$$\mathbf{p} = [1 \quad x \quad y]$$

Figure 8 *Pathological pattern with quadratic basis*
$$\mathbf{p} = [1 \ x \ y \ x^2/2 \ xy \ y^2/2]$$

In these examples, we chose $w(x_i, x) = w_{ref}(\|x_i - x\|/r)$ where w_{ref} is a bell shaped window function, $\| \cdot \|$ is the Euclidian norm L^2 and $r(x)$ is the distance from x to the $n_\mathbf{p} + 2$ neighbor node of x. The nodes selected at the point x are indicated by full dots. Figure 6 presents a well-conditioned case with a linear basis for a random distribution of nodes. Figure 7 and Figure 8 show particular cases where matrix A becomes singular, respectively with collinear points with a linear basis or with co-circular points with a quadratic basis. These pathological situations are the limit cases in which the approximation cannot be performed. The patterns close to these singular ones may lead to ill-conditioned matrices and therefore spoil the convergence. We note that these results do not depend on the choice of the weighting function and that they are similar to those obtained by (Syczewski and Tribillo, 1981) for a finite difference scheme on irregular mesh.

Expression (30) shows that when a linear polynomial basis $\mathbf{p}$ is used, the convergence of the function approximation is quadratic and the convergence of the diffuse derivative is linear. The advantage of MLS is a better control of the continuity properties provided by an appropriate choice of weight reference function w_{ref}. When a quadratic base $\mathbf{p}$ is used, a cubic convergence of the function together with a quadratic convergence of the first derivative and a linear convergence of the second derivative is obtained. This last property is mandatory when implementing a second order finite difference scheme commonly used in computational mechanics. However, the number of nodes involved in the computation and consequently the bandwidth of the resulting global system depends on the number of terms in $\mathbf{p}$ and augments with its degree. Moreover, in order to preserve the local support of the approximation, it is interesting to keep the degree of $\mathbf{p}$ reduced. Thus, a linear $\mathbf{p}$ will be frequently used in Galerkin-like formulations while a quadratic $\mathbf{p}$ is necessary in GFDM.

2.6. Stability of the numerical scheme

Figure 9 and Figure 10 illustrate the behavior of the condition number of the matrix $\mathbf{A}$ for the three different formulations of the MLS criterion: $J_x(\mathbf{a})$, $J_x(\alpha)$ and $J_x(\beta)$

as introduced in the expressions (3), (17) and (24). The polynomial basis chosen for this example is linear. The evaluation points are located on a regular grid covering the 1D domain $(-5, 5)$ discretized with equidistant nodes perturbed randomly by 30%. Three closest nodes are taken into account at each evaluation point. The Schwartz window reference function is used. The condition number is defined as the ratio of the largest to the smallest eigenvalue of the matrix $\mathbf{A}$. This value determines the precision of the MLS approximation coefficients obtained. Large values of the condition number indicate that the matrix is nearly singular.

Figure 9 shows the variation of the condition number over the domain for the three criteria and a ten-node discretization. $J_x(\mathbf{a})$ and $J_x(\alpha)$ give similar conditioning in the vicinity of the origin of the coordinate system chosen here at the center of the domain. $J_x(\beta)$ is slightly worse than $J_x(\mathbf{a})$ and $J_x(\alpha)$, but its condition number is always lower then 100 and is largely acceptable. When the distance from the origin increases, we observe an important degradation of $J_x(\mathbf{a})$ performance while $J_x(\alpha)$ and $J_x(\beta)$ give roughly bounded and constant conditioning. Random nodal positions result in irregular oscillation of the curves.

Figure 10 illustrates the case of a progressively refined set of nodes. Three neighboring nodes are again taken at each evaluation point and consequently, the size of the domains of influence decreases. We analyze the maximal value of the condition number at each evaluation point. We observe that $J_x(\alpha)$ is always better than $J_x(\mathbf{a})$ and differs by 2 orders of magnitude. However, both formulations gradually degenerate when the number of nodes increases. The dimensionless formulation

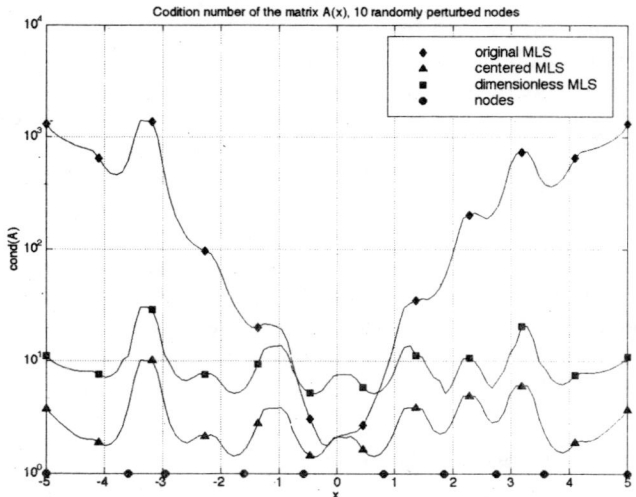

Figure 9 *Distribution of the condition number of the matrix A over a 1D domain with 10 randomly perturbed nodes. $J_x(\mathbf{a})$, $J_x(\alpha)$ and $J_x(\beta)$ formulations*

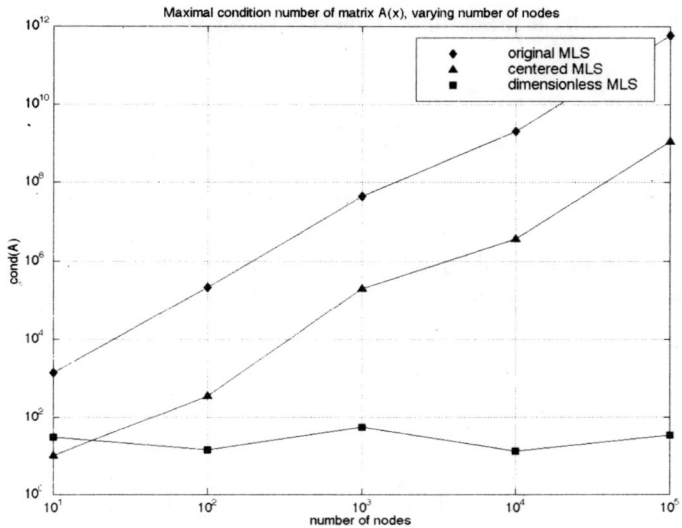

Figure 10 *Maximal condition number of the matrix A over a 1D domain with increasing density of randomly perturbed nodes. $J_x(\mathbf{a})$, $J_x(\alpha)$ and $J_x(\beta)$ formulations*

$J_x(\beta)$, while slightly worse than $J_x(\alpha)$ for very low numbers of nodes, is always well conditioned, independently of the nodal density. As in the previous figure, the slightly nonlinear character of lines is due to the random nodal positions.

Similar behavior is observed in 2D and 3D as well as for higher order polynomial basis. The use of the dimensionless formulation $J_x(\beta)$ is therefore mandatory when performing convergence studies and is strongly recommended in practical programming.

3. MLS shape functions

In the vocabulary of the finite element method, we may identify the coefficients α as the MLS equivalent of the shape functions.

In the context of the interpretation (21) we may rewrite relation (19) as

$$
\left\{
\begin{array}{c}
u_{app}(x) \\[4pt]
\dfrac{\delta u_{app}}{\delta x}(x) \\[4pt]
\vdots \\[4pt]
\dfrac{\delta u_{app}^k}{\delta x^k}(x)
\end{array}
\right\} = \mathbf{A}^{-1}\mathbf{B}\{u_i\}
\tag{31}
$$

or, alternatively with the usual finite element notation

$$u_{app}(\mathbf{x}) = \sum_i N_i(\mathbf{x})u_i$$

$$\frac{\delta u_{app}^l}{\delta x^l}(\mathbf{x}) = \sum_i \frac{\delta^l N_i(\mathbf{x})}{\delta x^l}u_i \tag{32}$$

All the shape functions N_i along with their diffuse derivatives are then expressed in a compact form as subsequent columns of a matrix resulting from product $\mathbf{A}^{-1}\mathbf{B}$

$$\begin{vmatrix} \mathbf{N}^T \\ \dfrac{\delta \mathbf{N}^T}{\delta x} \\ \dfrac{\delta \mathbf{N}^T}{\delta y} \\ \vdots \end{vmatrix} = \mathbf{A}^{-1}\mathbf{B} \tag{33}$$

The computational cost involved depends primarily on the inversion of matrix $\mathbf{A}$ which does not need to be performed explicitly. LU decomposition can be used instead (Belytchko *et al.*, 1996). The computation of the diffuse derivatives of the shape functions can be performed with no significant extra cost.

3.1. *Consistency conditions*

The consistency conditions, necessary for quadratic convergence of $u_{app}(x) = \sum_i N_i(x)u_i$ can be expressed as

$$\sum_i N_i = 1$$

$$\sum_i N_i(x_i - x) = 0 \tag{34}$$

The cubic convergence of u_{app} also requires

$$\sum_i N_i(x_i - x)^2 = 0 \tag{35}$$

The three consistency constraints (34), (35) can be presented compactly in a single matrix condition

$$\mathbf{P}\mathbf{N} = \mathbf{e}^1 \tag{36}$$

with

$$\mathbf{e}^1 = \begin{Bmatrix} 1 \\ 0 \\ 0 \end{Bmatrix}, \quad \mathbf{e}^2 = \begin{Bmatrix} 0 \\ 1 \\ 0 \end{Bmatrix}, \quad \text{etc}\ldots \tag{37}$$

and

$$\mathbf{P} = \begin{vmatrix} 1 & \cdots & 1 \\ x_1 - x & \cdots & x_n - x \\ \frac{1}{2}(x_1 - x)^2 & \cdots & \frac{1}{2}(x_n - x)^2 \end{vmatrix} \tag{38}$$

Linear convergence of $(du_{app}/dx)(x) = \sum_i N_i'(x)u_i$ requires

$$\sum_i N_i' = 0$$

$$\sum_i N_i'(x_i - x) = 1 \tag{39}$$

and for quadratic convergence we also have to satisfy

$$\sum_i N_i'(x_i - x)^2 = 0 \tag{40}$$

Thus, by analogy to (36) and using (37), (38) we obtain

$$\mathbf{PN}' = \mathbf{e}^2 \tag{41}$$

These properties, well known in finite elements under the name of consistency conditions, are necessary for the convergence of a variational formulation based on first derivatives, such as finite elements or diffuse elements.

In the collocation formulations based on second derivatives, the linear convergence of the terms $\frac{d^2 u_{app}}{dx^2}(x) = \sum_i N_i''(x)u_i$ appearing in the equilibrium equations, requires:

$$\sum_i N_i'' = 0$$

$$\sum_i N_i''(x_i - x) = 0$$

$$\sum_i N_i'' \frac{(x_i - x)^2}{2} = 1 \tag{42}$$

or

$$\mathbf{PN}'' = \mathbf{e}^3 \tag{43}$$

Conditions (36), (41) and (43) are automatically satisfied when deriving shape functions based on a complete quadratic polynomial basis by MLS or by GFDM. In the next section, we define a way to construct the shape functions directly from the properties required.

3.2. *Consistency based approach for shape functions determination*

We now introduce an alternative technique of shape function construction explicitly based on the desired consistency conditions. This approach presents several advantages both from the computational and from the formal point of view. The algorithm obtained is efficient and some supplementary conditions, different from consistency, can also be introduced.

Let us consider an evaluation point x and a set of associated nodes $\{x_1, \ldots, x_n\}$ "close to" the point x. We note

$$\mathbf{W} = \begin{vmatrix} w_1 & & 0 \\ & \ddots & \\ 0 & & w_n \end{vmatrix} \tag{44}$$

We introduce the objective function

$$J(\mathbf{N}) = \frac{1}{2}\mathbf{N}^T\mathbf{W}^{-1}\mathbf{N} \tag{45}$$

and we look for functions $\mathbf{N}$ which are solutions of $Min(J(\mathbf{N}))$ subjected to the first order consistency constraints (36).

The associated Lagrangian is

$$L(\mathbf{N}, \lambda) = \frac{1}{2}\mathbf{N}^T\mathbf{W}^{-1}\mathbf{N} + \lambda^T(\mathbf{PN} - \mathbf{e}^1) \tag{46}$$

and the optimality conditions are

$$\mathbf{PN} - \mathbf{e}^1 = 0$$
$$\mathbf{N}^T\mathbf{W}^{-1} + \lambda^T\mathbf{P} = 0 \tag{47}$$

leading to the following linear system

$$\begin{vmatrix} \mathbf{W}^{-1} & \mathbf{P}^T \\ \mathbf{P} & 0 \end{vmatrix} \begin{Bmatrix} \mathbf{N} \\ \lambda \end{Bmatrix} = \begin{Bmatrix} 0 \\ \mathbf{e}^1 \end{Bmatrix} \tag{48}$$

The solution of (48) is given by $\mathbf{N} = -\mathbf{WP}^T\lambda$, so $\mathbf{P}(-\mathbf{WP}^T\lambda) = \mathbf{e}^1$ and

$$\lambda = -(\mathbf{PWP}^T)^{-1}\mathbf{e}^1 \tag{49}$$

and finally

$$\mathbf{N}^T = \mathbf{e}^{1^T}(\mathbf{PWP}^T)^{-1}\mathbf{PW} \tag{50}$$

where we recognize the matrices (20)

$$\mathbf{A} = \mathbf{PWP}^T$$
$$\mathbf{B} = \mathbf{PW}$$

(51)

Functions $\mathbf{N}$ in the expression (50) correspond obviously to the moving least squares shape functions.

Moreover, the first derivatives of the shape functions $\mathbf{N}'$ are solutions of $Min(J(\mathbf{N}'))$ under the constraint (41). The second derivative, $\mathbf{N}''$ is a solution of $Min(J(\mathbf{N}''))$ under the constraint (43), leading to the following expressions for the first and to the second diffuse derivatives of the shape functions

$$\mathbf{N}'^T = \mathbf{e}^{2^T} (\mathbf{PWP}^T)^{-1} \mathbf{PW}$$
$$\mathbf{N}''^T = \mathbf{e}^{3^T} (\mathbf{PWP}^T)^{-1} \mathbf{PW}$$

(52)

We establish in this way that

$$N_i'(x) \equiv \frac{\delta N_i}{\delta x}(x)$$
$$N_i''(x) \equiv \frac{\delta^2 N_i}{\delta x^2}(x)$$

(53)

Finally, we find again formula (33)

$$\begin{vmatrix} N_1(x) & \cdots & N_n(x) \\ \dfrac{\delta N_1}{\delta x}(x) & \cdots & \dfrac{\delta N_n}{\delta x}(x) \\ \dfrac{\delta^2 N_1}{\delta x^2} N_1(x) & \cdots & \dfrac{\delta^2 N_n}{\delta x^2}(x) \end{vmatrix} = \mathbf{A}(x)^{-1}\mathbf{B}$$

(54)

It is important to note that

- the subsequent derivatives of the functions N are obtained here as a result of the minimization of criterion (45) subjected to different consistency constraints (36), (41) and (43),
- we have shown that these "consistency based" derivatives correspond to the diffuse derivatives. It implies that the diffuse derivatives are sufficient for the convergence of the solution of PDEs.

Other constraints that consistency demands may be applied. Therefore, this alternative presentation of the shape functions based on explicit constraints provides a powerful way to handle varied optimization constraints. This includes equality constraints such as incompressibility (Huerta *et al.*, 2002) and an inequality constraint such as plastic admissibility (Breitkopf *et al.*, 2001). The $J(\mathbf{N})$ formulation was also used for the development of an efficient, well-conditioned algorithm for the shape functions evaluation, without an explicit inversion of the matrix $\mathbf{A}$ (Breitkopf *et al.*, 2000).

3.3. *Explicit form of the shape functions*

A further insight into the MLS methodology can be given by developing an explicit shape functions formulation. When the linear consistency constraints are required, the task leads to an inversion of $2 * 2$, $3 * 3$ and $4 * 4$ matrices respectively for the 1D, 2D and 3D cases.

3.3.1. Shape functions and their derivatives in 1D

The domains of influence of nodes are chosen in order to provide a 3-node connectivity, respectively x_1, x_2, x_3, at each evaluation point x. The weights are denoted as w_1, w_2, w_3.

Linear consistency constraints (36) are obtained with the matrix $\mathbf{P}$

$$\mathbf{P} = \begin{vmatrix} 1 & 1 & 1 \\ x_1 - x & x_2 - x & x_3 - x \end{vmatrix} \tag{55}$$

We then have the diagonal weight matrix

$$\mathbf{W} = \begin{vmatrix} w_1 & 0 & 0 \\ 0 & w_2 & 0 \\ 0 & 0 & w_3 \end{vmatrix} \tag{56}$$

and performing the necessary algebra we get the following explicit expressions of the three shape functions

$$
\begin{aligned}
N_1 &= \frac{w_1 w_2 (x - x_2)(x_1 - x_2) + w_1 w_3 (x - x_3)(x_1 - x_3)}{w_1 w_2 (x_1 - x_2)^2 + w_3 w_2 (x_3 - x_2)^2 + w_1 w_3 (x_1 - x_3)^2} \\
N_2 &= \frac{w_2 w_1 (x - x_1)(x_2 - x_1) + w_2 w_3 (x - x_3)(x_2 - x_3)}{w_1 w_2 (x_1 - x_2)^2 + w_3 w_2 (x_3 - x_2)^2 + w_1 w_3 (x_1 - x_3)^2} \\
N_3 &= \frac{w_3 w_1 (x - x_1)(x_3 - x_1) + w_3 w_2 (x - x_2)(x_3 - x_2)}{w_1 w_2 (x_1 - x_2)^2 + w_3 w_2 (x_3 - x_2)^2 + w_1 w_3 (x_1 - x_3)^2}
\end{aligned}
\tag{57}
$$

The recursive approach for an arbitrary number n of nodes in 1D gives a general expression of the shape functions in the form

$$N_i = \frac{\sum_{j\neq1} w_i w_j (x - x_j)(x_i - x_j)}{d}$$

$$d = \sum_{i=1,n-1} \sum_{j=i+1,n} w_i w_j (x_i - x_j)^2$$

(58)

and the "full" x derivative is given by

$$\frac{dN_i}{dx} = \frac{1}{d} \sum_{j\neq1} w_i w_j (x_i - x_j) + \frac{1}{d} \sum_{j\neq1} (w_{i,x} w_j + w_i w_{j,x})(x - x_j)(x_i - x_j) - N_i$$

$$d' = \sum_{i=1,n-1} \sum_{j=i+1,n} (w_{i,x} w_j + w_i w_{j,x})(x_i - x_j)^2$$

(59)

where the first term corresponds to the diffuse derivative $w_{i,x} = 0$

$$\frac{\delta N_i}{\delta x} = \frac{1}{d} \sum_{j\neq1} w_i w_j (x_i - x_j)$$

(60)

One may also note that when $w_i = 0$ for $i > 2$ or when the number of nodes is equal to 2, the approximation degenerates as expected to the well known finite element shape functions:

$$N_1 = \frac{x - x_2}{x_1 - x_2}, \quad \frac{dN_1}{dx} = \frac{\delta N_1}{\delta x} = \frac{1}{x_1 - x_2}$$

$$N_2 = \frac{x_1 - x}{x_1 - x_2}, \quad \frac{dN_2}{dx} = \frac{\delta N_2}{\delta x} = \frac{-1}{x_1 - x_2}$$

(61)

3.3.2. Shape functions and their derivatives in 2D

When taking a 4 node neighborhood x_1, x_2, x_3, x_4 at each evaluation point x in 2D with corresponding weights w_1, w_2, w_3, w_4 together with linear consistency constraints we obtain the first shape function

$$N_1 = \frac{1}{d}(w_1 w_2 w_3(-x_2 y + x_3 y + x y_2 - x_3 y_2 - x y_3 + x_2 y_3)$$
$$\times (-x_2 y_1 + x_3 y_1 + x_1 y_2 - x_3 y_2 - x_1 y_3)$$
$$\times w_1 w_2 w_4(-x_2 y + x_4 y + x y_2 - x_4 y_2 - x y_4 + x_2 y_4)$$
$$\times (-x_2 y_1 + x_4 y_1 + x_1 y_2 - x_4 y_2 - x_1 y_4 + x_2)$$
$$\times w_1 w_3 w_4(-x_3 y + x_4 y + x y_3 - x_4 y_3 - x y_4 + x_3 y_4)$$
$$\times (-x_3 y_1 + x_4 y_1 + x_1 y_3 - x_4 y_3 - x_1 y_4 + x_3)) \tag{62}$$

The other shape functions are given by similar expressions, which can be written under the general form

$$N_1 = \frac{w_i}{d} \sum_{j \neq i} \sum_{k > j, k \neq i} w_j w_k \Theta_{2D}(\mathbf{x}, \mathbf{x_j}, \mathbf{x_k}) \Theta_{2D}(\mathbf{x_i}, \mathbf{x_j}, \mathbf{x_k})$$

$$d = \sum_{i=1, n-2} \sum_{j=i+1, n-1} \sum_{k=j, n} w_i w_j w_k (\Theta_{2D}(\mathbf{x_i}, \mathbf{x_j}, \mathbf{x_k}))^2 \tag{63}$$

where

$$\Theta_{2D}(\mathbf{x_i}, \mathbf{x_j}, \mathbf{x_k}) = -x_j y_i + x_k y_i + x_i y_j - x_k y_j - x_i y_k + x_j y_k \tag{64}$$

and the diffuse derivatives are

$$\frac{\delta N_i}{\delta x} = \frac{w_i}{d} \sum_{j \neq i} \sum_{k > j, k \neq i} w_j w_k (y_j - y_k) \Theta_{2D}(\mathbf{x_i}, \mathbf{x_j}, \mathbf{x_k})$$

$$\frac{\delta N_i}{\delta y} = \frac{w_i}{d} \sum_{j \neq i} \sum_{k > j, k \neq i} w_j w_k (x_k - x_j) \Theta_{2D}(\mathbf{x_i}, \mathbf{x_j}, \mathbf{x_k}) \tag{65}$$

Full derivatives can also be easily obtained by differentiating expression (63) or by formal differentiation of the computer code.

3.3.3. Shape functions and their derivatives in 3D

For 3D shape functions

$$N_i = \frac{w_i}{d} \sum_{j \neq i} \sum_{j < k, k \neq i} \sum_{k < l, l \neq i} w_j w_k w_l \Theta_{3D}(\mathbf{x}, \mathbf{x_j}, \mathbf{x_k}, \mathbf{x_l}) \Theta_{3D}(\mathbf{x_i}, \mathbf{x_j}, \mathbf{x_k}, \mathbf{x_l})$$

$$d = \sum_{i=1, n-3} \sum_{i < j < n-2} \sum_{j < k < n-1} \sum_{k < l < n} w_i w_j w_k w_l (\Theta_{3D}(\mathbf{x_i}, \mathbf{x_j}, \mathbf{x_k}, \mathbf{x_l}))^2 \tag{66}$$

with

$$\Theta_{3D}(\mathbf{x_i}, \mathbf{x_j}, \mathbf{x_k}, \mathbf{x_l}) = z_i \Theta_{2D}(\mathbf{x}_j, \mathbf{x}_k, \mathbf{x}_l) + z_j \Theta_{2D}(\mathbf{x}_i, \mathbf{x}_k, \mathbf{x}_l)$$
$$+ z_k \Theta_{2D}(\mathbf{x}_i, \mathbf{x}_j, \mathbf{x}_l) + z_l \Theta_{2D}(\mathbf{x}_i, \mathbf{x}_j, \mathbf{x}_k) \tag{67}$$

The corresponding diffuse derivatives are

$$\frac{\delta N_i}{\delta x} = \frac{w_i}{d} \sum_{j \neq i} \sum_{j<k,k \neq i} \sum_{k<l,l \neq i} \Theta_{2D}\left(\left\{\begin{matrix} y_j \\ z_j \end{matrix}\right\}, \left\{\begin{matrix} y_k \\ z_k \end{matrix}\right\}, \left\{\begin{matrix} y_l \\ z_l \end{matrix}\right\}\right) \Theta_{3D}(\mathbf{x_i}, \mathbf{x_j}, \mathbf{x_k}, \mathbf{x_l})$$

$$\frac{\delta N_i}{\delta y} = \frac{w_i}{d} \sum_{j \neq i} \sum_{j<k,k \neq i} \sum_{k<l,l \neq i} \Theta_{2D}\left(\left\{\begin{matrix} y_j \\ z_j \end{matrix}\right\}, \left\{\begin{matrix} y_k \\ z_k \end{matrix}\right\}, \left\{\begin{matrix} y_l \\ z_l \end{matrix}\right\}\right) \Theta_{3D}(\mathbf{x_i}, \mathbf{x_j}, \mathbf{x_k}, \mathbf{x_l})$$

$$\frac{\delta N_i}{\delta z} = \frac{w_i}{d} \sum_{j \neq i} \sum_{j<k,k \neq i} \sum_{k<l,l \neq i} \Theta_{2D}\left(\left\{\begin{matrix} y_j \\ x_j \end{matrix}\right\}, \left\{\begin{matrix} y_k \\ x_k \end{matrix}\right\}, \left\{\begin{matrix} y_l \\ x_l \end{matrix}\right\}\right) \Theta_{3D}(\mathbf{x_i}, \mathbf{x_j}, \mathbf{x_k}, \mathbf{x_l})$$

$$\tag{68}$$

We observe that the practical use of these explicit forms of the shape functions is limited by the cost of their evaluation. The cost of the numerical algorithm is linear with respect to the number of nodes in the neighborhood of $\mathbf{x}$. However, the number of operations involved in a straightforward evaluation of the latter explicit formula for n nodes and k constraints is proportional to

$$C_n^k = \frac{n!}{(n-k)!k!} \tag{69}$$

resulting in an $0(k)$ complexity. The formal expressions, which can still be applied for linear consistency constraints, become fairly complex when quadratic consistency is required. The cost of an explicit inversion of a $6 * 6$ matrix in 2D or of a $10 * 10$ matrix in a 3D becomes prohibitive.

The knowledge of the explicit forms is however important when considering issues such as continuity, the implementation of boundary conditions or integration strategies in variational formulations.

4. Interpolating MLS

In a general case, the MLS approximation does not interpolate data. In order to enforce the interpolating condition (4) at a node i, we separate the ith term in the optimality condition (18) for $J_x(\alpha)$

$$\lambda \mathbf{q}^T (x_i - x) + \sum_{j \neq i} (\mathbf{q}^T (x_j - x)\alpha - u_j) w_j(x_j, x) \mathbf{q}^T (x_j - x) = 0 \tag{70}$$

where

$$\lambda = w_i(\mathbf{q}^T(x_i - x)\alpha - u_i) \tag{71}$$

In a matrix form, equations (70) and (71) become

$$\begin{bmatrix} \mathbf{A}_i & \mathbf{q}(x_i - x) \\ \mathbf{q}(x_i - x)^T & -1/w_i \end{bmatrix} \begin{Bmatrix} \alpha \\ \lambda \end{Bmatrix} = \begin{Bmatrix} \mathbf{b}_i \\ u_i \end{Bmatrix} \tag{72}$$

with $\mathbf{A}_i = \sum_{j \neq i} \mathbf{q}(x_j - x)^T w_j \mathbf{q}(x_j - x)$ and $\mathbf{b}_i = \sum_{j \neq i} w_j u_j \mathbf{q}(x_j - x)$. When the weight function is singular at node i (5), then system (72) tends uniformly to

$$\begin{bmatrix} \mathbf{A}_i & \mathbf{e}^1 \\ \mathbf{e}^{1^T} & 0 \end{bmatrix} \begin{Bmatrix} \alpha \\ \lambda \end{Bmatrix} = \begin{Bmatrix} \mathbf{b}_i \\ u_i \end{Bmatrix} \tag{73}$$

which is precisely the optimality condition (18) for $J_x(\alpha)$ under the constraint (4) written as

$$u_{app}(x_i) = \mathbf{q}^T(x_i - x_i)\alpha = \alpha_0 = u_{ex}(x_i) \tag{74}$$

where λ appears as the Lagrange multiplier.

It is interesting to examine the properties of the diffuse (6) and full (7) derivatives of the MLS interpolation at node x_i. When deriving (72), we obtain

$$\begin{bmatrix} \mathbf{A}_{i,x} & \mathbf{q}_{i,x} \\ \mathbf{q}_{i,x}^T & w_{i,x}/w_i^2 \end{bmatrix} \begin{Bmatrix} \alpha \\ \lambda \end{Bmatrix} + \begin{bmatrix} \mathbf{A}_i & \mathbf{q}_i \\ \mathbf{q}_i^T & -1/w_i \end{bmatrix} \begin{Bmatrix} \alpha_{,x} \\ \lambda_{,x} \end{Bmatrix} = \begin{Bmatrix} \mathbf{b}_{i,x} \\ 0 \end{Bmatrix} \tag{75}$$

We observe that when $x \to x_i$ then $\mathbf{q}(x_i - x) \to \mathbf{e}^1$ and $\mathbf{q}_{,x}(x_i - x) \to -\mathbf{e}^2$, thus $\mathbf{q}_i^T(d\alpha/dx) \to d\alpha_1/dx = du/dx$ and $\mathbf{q}_{i,x}^T\alpha \to -\alpha_2 = -\delta u/\delta x$. The last line of the matrix form (75) becomes

$$-\frac{\delta u}{\delta x} + \lambda\frac{w_{i,x}}{w_i^2} + \frac{du}{dx} - \frac{1}{w_i}\lambda_{,x} = 0 \tag{76}$$

For singular weights, when the derivative of the reference weight function w_i is bounded, the detailed analysis of the system (73) shows that the second and fourth terms in (76) disappear for $x \to x_i$ (5). We then have

$$\lim_{x \to x_i} \frac{\delta u}{\delta x} = \frac{du}{dx} \tag{77}$$

This property means that, in the interpolating version of the MLS, the diffuse derivative is equal to the full derivative at the node. This feature is important in the meshfree methods based on nodal integration schemes or in nodal collocation methods.

4.1. *Singular weights*

The singular weights can be obtained by scaling the original weight functions in order to give a unit value at a node $w_i(x_i) = 1$ and then by applying the following substitution

$$[w(x_i, x)] \rightarrow \left[\frac{w(x_i, x)}{1 - w(x_i, x)} \right] \qquad (78)$$

Interpolating shape functions are then obtained by minimization of any of the proposed criteria $J_x(\mathbf{a})$, $J_x(\alpha)$, $J_x(\beta)$, $J(\mathbf{N})$ with modified weights at any evaluation point $x \neq x_i$.

When an evaluation point is located at a node i, the modified weight function becomes singular. Nevertheless, the shape functions are known at the node without computation and are given by the interpolation condition:

$$N_i(x = x_j) = \delta_i^j \qquad (79)$$

However, when the evaluation point is "close to" but not exactly "at" the node, then the use of singular weights becomes uncomfortable. Several methods may be applied but in practice it is sufficient to limit the growth of the $\tilde{w}_i$ by taking

$$\tilde{w}_i = \frac{w_i}{1 + \varepsilon - w_i} \qquad (80)$$

where ε is a "small" value. It is then not necessary to distinguish between the two separate cases (78), (79) and the resulting approximation continuity is limited only by the continuity of the reference weight function.

Such "near singular weights" may result however in an ill-conditioning of the algorithm.

4.2. *Interpolation with non-singular weights*

A different strategy based on a Shepard interpolation can be used in order to obtain an interpolating MLS approximation. We notice first that the substitution

$$[w] \rightarrow \lambda[w] \qquad (81)$$

does not modify the solution of the optimality system for any $\lambda \neq 0$. Let us now introduce the scaled weights $\tilde{w}_j$

$$S(x) = \sum_i w(x_i, x), \quad \tilde{w}_j = \frac{1}{S(x)} w(x_j, x) \qquad (82)$$

In the neighborhood of node $\mathbf{x}_j$ the modified weight function $\tilde{w}_j$ may be then written without singularity

$$\tilde{w}(\mathbf{x}_j, \mathbf{x}) = \frac{w_j}{w_j + (1 - w_j)\sum_{i \neq j}[w_i/(1 - w_i)]} \tag{83}$$

and $\tilde{w}(\mathbf{x}_j, \mathbf{x}_j) = 1$.

In the neighborhood of a node $\mathbf{x}_k, k \neq j$, expression (83) becomes singular as $w_k \to 1$. However, the weight function $\tilde{w}_j$ may now be computed from the expression (83) reformulated in the following way

$$\tilde{w}(\mathbf{x}_j, \mathbf{x}) = \frac{w_j(1 - w_k)}{(1 - w_j)\left(w_k + (1 - w_k)\sum_{i \neq k}[w_i/(1 - w_i)]\right)} \tag{84}$$

and we see that $\tilde{w}(\mathbf{x}_j, \mathbf{x}_k) = 0, k \neq j$.

These modified weights have the following properties

$$\tilde{w}(x_i, x) \in [0, 1], \quad \sum_i \tilde{w}(x_i, x) = 1, \quad \tilde{w}(x_i, x_j) = \delta_{ij} \tag{85}$$

Figure 11 and Figure 12 illustrate the normalized interpolating weighting functions, derived from the reference weights (8) and (9) for a three-node configuration.

The weights obtained by this procedure are not singular, taking a unit value at their reference node and vanishing at other nodes.

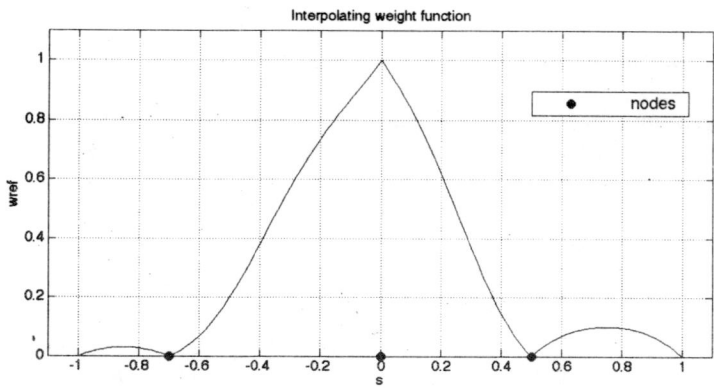

Figure 11 *Nonsingular interpolating weights derived from the hat reference weight*

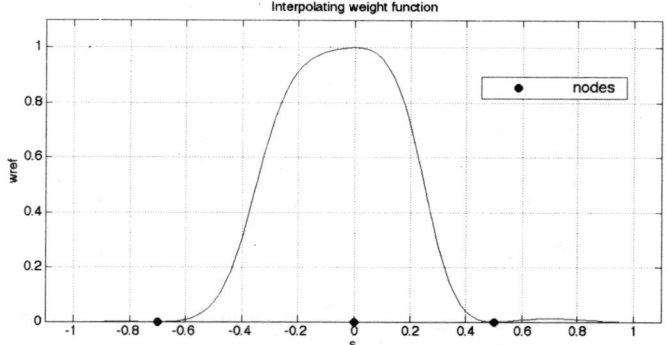

Figure 12 *Nonsingular interpolating weights derived from the spline reference weight*

5. Diffuse elements

We have chosen to develop a meshfree method which may be implemented at a minimal cost within a standard finite element software framework. By analogy with finite elements, we define a "diffuse element" which can be used in the assembly procedure for the global matrix. The diffuse element is identified by a list of nodes with non-zero contributions. From the geometrical point of view, a diffuse element corresponds to the intersections of the domains of influence of connected nodes. MLS shape functions are used instead of their finite element equivalents in order to obtain elementary matrices and vectors through a process of numerical integration. The issue of definition of domains of influence is tightly coupled to that of the precision of numerical integration (Dolbow and Belytschko, 1999). We extend here their approach in order to reduce the number of integration cells.

This approach is not the only way to construct a meshfree method and alternative approaches which do not use a background grid for numerical integration have been proposed. These methods include the "truly meshless" techniques (Lin and Atluri, 2000, De and Bathe, 2000) and nodal integration methods (Beissel and Belytschko, 1996, Bonet and Lok, 1999, Chen *et al.*, 2002).

The implementation of the "diffuse element" approach is straightforward. However, several questions which do not arise in the finite element method, have still to be solved. These include primarily the issues of

- domain decomposition,
- numerical integration schemes,
- essential boundary conditions,
- patch test.

In the finite element method, the continuum Ω is divided into a finite number (say E) of open disjoint subregions – finite elements $\{\Omega_e, e = 1, 2, \ldots, E\}$ such that for **interior**$(\Omega_e) \cap$ **interior**$(\Omega_f) = \emptyset$ for $e \neq f$. The finite element interpolation is

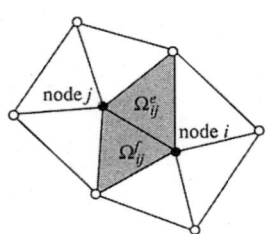

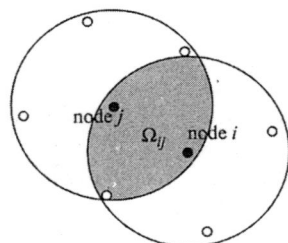

Figure 13 *Finite elements*   **Figure 14** *Diffuse elements, r_i*
strategy
Example of integration cells for the term (i, j) of the global system

defined locally in each element. The finite elements are also used as integration cells for numerical evaluation of the global integrals over the domain Ω, generally using the Gauss-Legendre scheme. The contribution (i, j) to the global linear system is then assembled from the set of elements sharing the nodes i and j (Figure 13).

In the diffuse element method, the (i, j) term of the global system is integrated over the intersection of domains of influence of nodes i and j (Figure 14). The evaluation of integrals over Ω is however less obvious as the domains of influence overlap in general in an irregular manner. Moreover, the domains do not respect the boundary. Thereafter, the integration scheme depends on the strategy in which we define the influence domains of the nodes (cf. section 2.2.3.).

$R(x)$ strategy guarantees the existence of the MLS approximation at each evaluation point belonging to Ω. Figure 15 illustrates the domains of influence for a regular grid of nodes with $k = 4$. The individual integration cells are then given by the Voronoï diagram of 4-th order polygons defined by the sets of points sharing the same list of 4 closest neighboring nodes. In the following figures, the nodes are denoted by thick dots.

We can notice that the shape of the integration domains is complex even if the grid of nodes is regular. Moreover, the generation of high order Voronoï diagrams is much more costly then the generation of a finite element mesh which only requires a first order Voronoï diagram.

The shapes of domains of influence of nodes is greatly simplified when using r_i strategy. However, when the L^2 norm is used for computing distances, the domains of influence are circular and the cost of computing individual contributions over all the intersections may still be prohibitive (Figure 16). In Figure 17 we show the domains of influence obtained using the L^∞ norm. An integration domain is given in this case by intersection of rectangular domains of influence of the nodes. For this reason, in the actual work we choose the L^∞ norm.

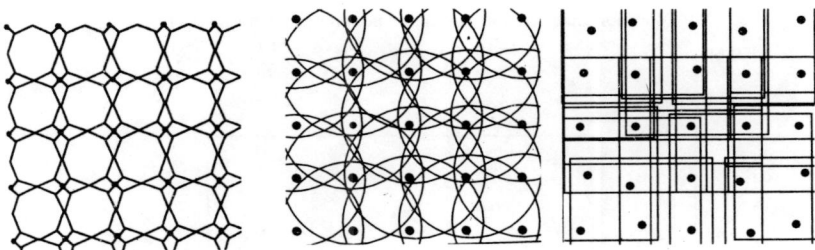

Figure 15 *R(x) strategy*   **Figure 16** *r_i strategy in*   **Figure 17** *r_i strategy and,*
                                            L^2 norm   *L^∞ norm*

Shapes of integration cells using various definitions of the radius of influence on a regular
(Figure 15, Figure 16) and on a randomly perturbed 2D grid (Figure 17)

Another difficulty when using r_i strategy consists in satisfying the two contradictory requirements:

– the domains of influence have to be big enough in order to guarantee the existence of the approximation at each point of Ω;
– the domains of influence should be as small as possible in order to limit the bandwidth of the resulting global system.

The second requirement governs also the accuracy of the approximation. It may be shown that these conditions are satisfied with the procedure illustrated in Figure 18. First, we build the first order Voronoï diagram and then, for each node we create the domain of influence as a rectangular envelope of Voronoï cells surrounding the cell to which the node belongs. This technique is sufficient for the linear basis $\mathbf{p} = \langle 1 \quad x \quad y \rangle$ as it connects at least 4 neighbors at each evaluation point.

We note that node "i" is not centered in the resulting domain of influence. We handle this situation in the following way. First, we introduce a C^1 continuous mappings $\xi(x)$ and $\eta(y)$ from asymmetric domain in 1D to a symmetric one along each axis (Figure 20).

Then, we define the 2D weight function as a tensor product of 1D weight functions $w(x, y) = w_{ref}(\xi)w_{ref}(\eta)$. Figure 21, Figure 22 and Figure 23 show a typical interpolating shape function and its derivatives over an asymmetric domain.

The common choice for numerical integration is the usual Gauss Legendre scheme. The integration points for rectangular domains are directly obtained by a linear mapping from the reference domain. In Figure 24, the solid tessels correspond to internal integration cells and the shaded ones belong to the boundary tessels. The

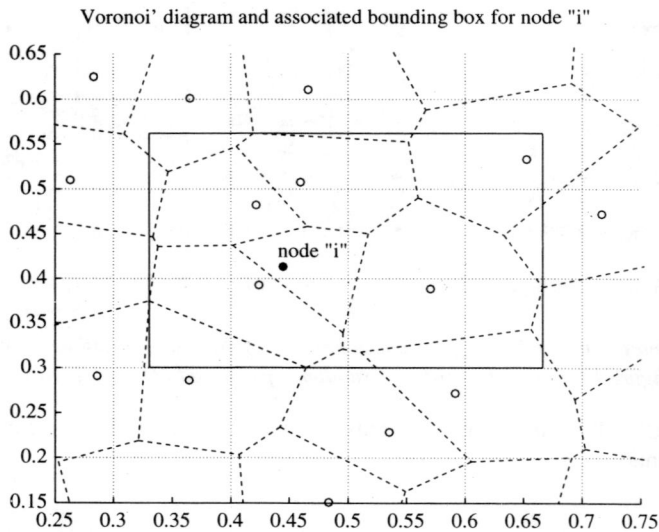

Figure 18 *Domain of influence defined as an envelope of surrounding Voronoï cells*

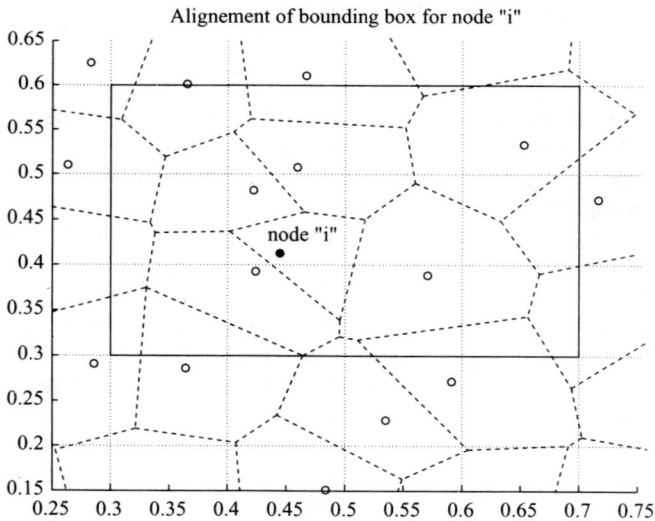

Figure 19 *Domain of influence expanded to fit the tessellation grid*

cells intersected by the boundary Γ have more complex forms and must be treated separately. For these cases, an isoparametric mapping can be used in a similar way as in a finite element context. Another solution consists in subdividing the boundary tessels into simpler shapes. Figure 25 provides the typical integration cells in 2D.

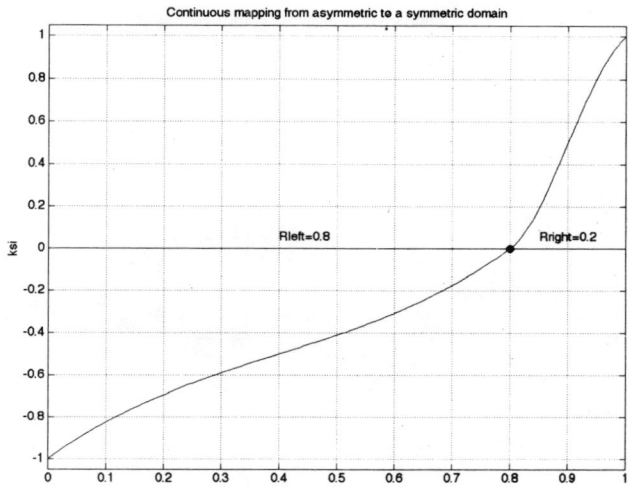

Figure 20 *Mapping used for nodes excentered in their domain of influence in 1D*

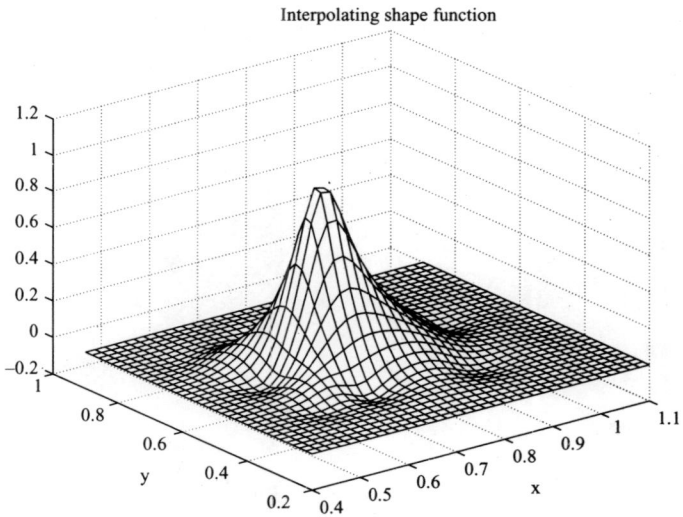

Figure 21 *Interpolating shape function over an asymmetric domain*

6. Integration scheme

6.1. *Patch test*

In the "patch test", a linear elasticity problem is solved over a domain with the displacements prescribed along all outside boundaries by a linear function

Interpolating shape function *x* derivative

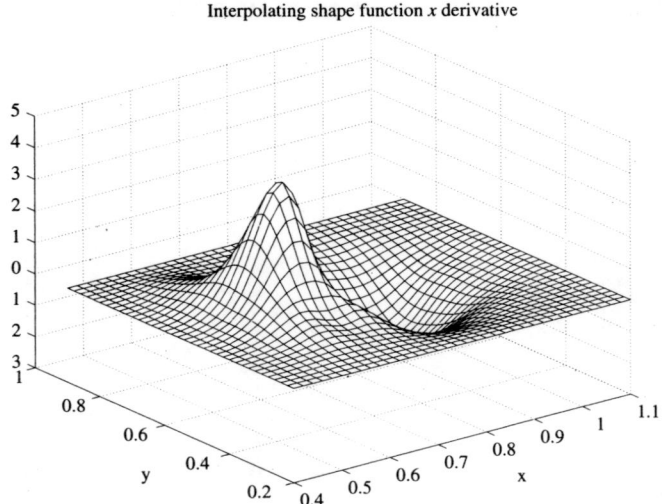

Figure 22 *Interpolating shape function x derivative over an asymmetric domain*

Interpolating shape function *y* derivative

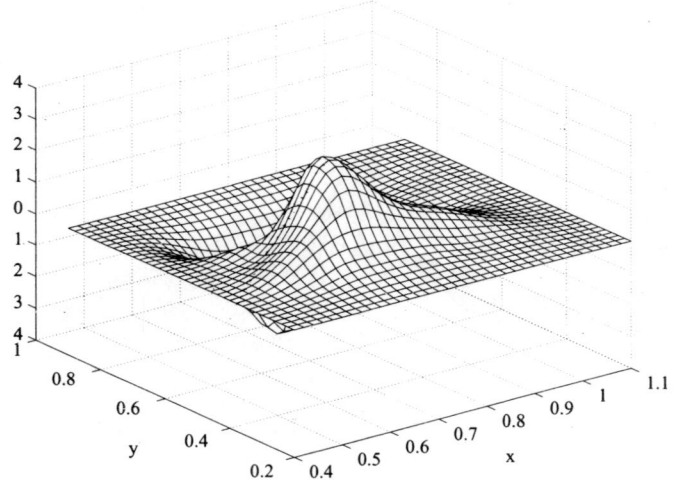

Figure 23 *Interpolating shape function y derivative over an asymmetric domain*

of the coordinates. The resulting strains and stresses in this case are constant. When a numerical method of solving the partial differential equations verifies this condition, we say that it satisfies the "patch test". This approach to verify the numerical formulation and the code itself is standard in the finite element

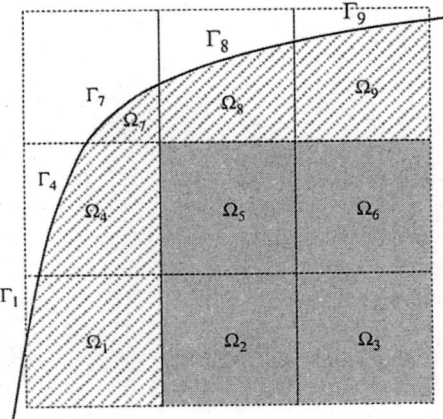

Figure 24 *Integration cells: internal (solid) and boundary (hatched)*

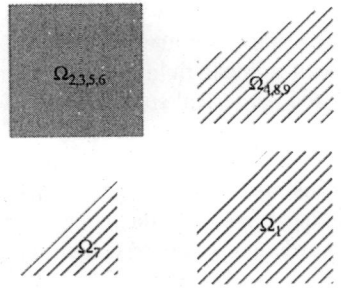

Figure 25 *Typical integration cells in 2D*

method. In the following section, we use our tessellation scheme along with the DEM and EFG formulations. Details of the reference test problem can be found in (Lu *et al.*, 1994). In the present work, the boundary conditions are enforced using the modified variational principle (Mukherjee and Mukherjee, 1997).

We define a domain Ω with boundary $\Gamma = \Gamma_N \cup \Gamma_D$. We study the Laplace equation

$$-\Delta u = f \quad \text{in } \Omega$$

$$u = \bar{u} \text{ on } \Gamma_D \quad \text{and} \quad \frac{\partial u}{\partial n} = \bar{q} \text{ on } \Gamma_N \tag{86}$$

where n is the outer normal defined on the boundary.

We use the extended variational formulation analogous to that given by (Lu *et al.*, 1994)

$$\int_\Omega \nabla u \nabla v \, d\Omega - \int_{\Gamma_D} \frac{\partial (uv)}{\partial n} d\Gamma = \int_\Omega f v \, d\Omega + \int_{\Gamma_N} \bar{q} v \, d\Gamma - \int_{\Gamma_D} \bar{u} \frac{\partial v}{\partial n} d\Gamma \quad (87)$$

$$\forall v \in \{v \in H^1(\Omega)/\Delta v \in L^2(\Omega)\}$$

The associated discrete system

$$\mathbf{Ku = F} \quad (88)$$

is obtained in the usual way with

$$K_{ij} = \int_\Omega \nabla N_i \nabla N_j \, d\Omega - \int_{\Gamma_D} \left(N_i \frac{\partial N_j}{\partial n} + N_j \frac{\partial N_i}{\partial n} \right) d\Gamma$$

$$F_i = \int_\Omega f(x) N_i(x) \, d\Omega + \int_{\Gamma_N} \bar{q} N_i \, d\Gamma - \int_{\Gamma_D} \bar{u} \frac{\partial N_i}{\partial n} d\Gamma \quad (89)$$

where N_i are the usual MLS shape functions.

The boundary conditions on both Neumann Γ_N and Dirichlet Γ_D parts of the boundary are associated with a linear field u_{ex}. In order to verify the patch test, we have to check whether the numerical solution procedure restitutes u_{ex} exactly inside Ω.

Figure 26 and Figure 27 illustrate the convergence of the patch test in 1D. Three nodes are used and we test the precision of the patch test versus the precision of the numerical integration. The orders of standard Gauss integration vary from 1 to 12. Different numbers of integration cells, with interpolating version of MLS shape functions, alternatively with full and with diffuse derivative, are used.

The following conclusions may be stated:

− in both cases the patch test is not a priori satisfied;
− when the "full" derivative is used, the patch test converges independently of the tessellation density, when refining the numerical integration; the error is however relatively important even for a high number of Gauss points; further refinement of the integration scheme leads to numerical errors;
− the diffuse derivative performs poorly, independently of the tessellation density and of the number of Gauss points.

The first two points can be easily explained when considering the explicit expressions of the shape functions established in paragraph 3.3. As opposed to the (simplest) case of finite elements, MLS shape functions do not have a polynomial form. In fact, when the weights w are given by spline functions, MLS shape functions are rational and their order is both defined by the number of connected nodes

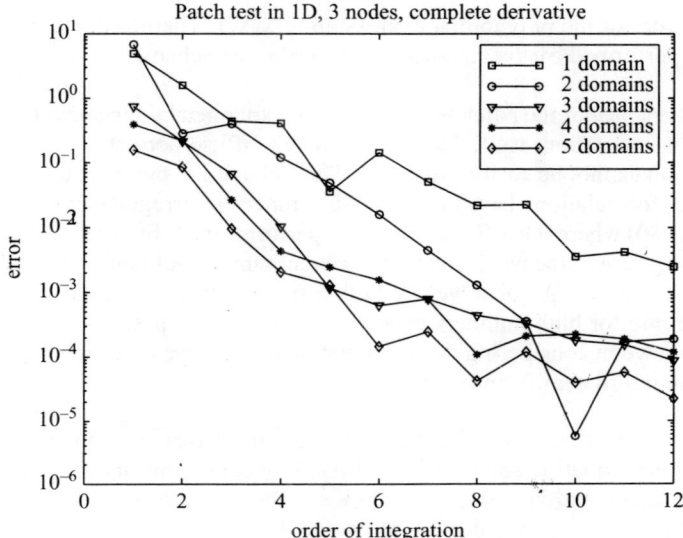

Figure 26 *Full derivative*

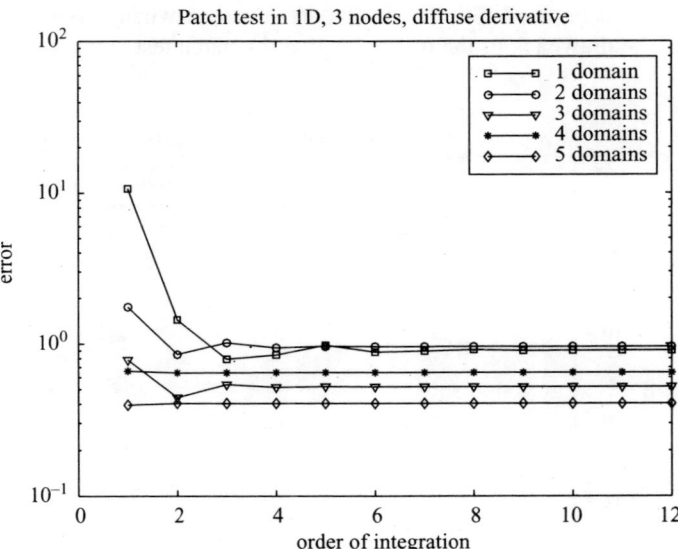

Figure 27 *Diffuse derivative*
*Precision of 3-node patch test in 1D for varying orders of standard Gauss integration and
different numbers of integration cells*

and the order of the polynomial expression $w(x, x_i)$. Therefore, the integration is not well performed by the classical Gauss-Legendre scheme.

The convergence of the patch test can be then explained by the convergence of the numerical integration itself. The failure of the diffuse derivative in a variational formulation cannot be easily explained. This behavior is opposite to that observed for strong formulations based on finite differences on irregular grids (Liszka and Orkisz, 1980) where the diffuse derivation performs well. In Figure 28, we analyze the convergence of the patch test with different numbers of nodes and with full and diffuse derivatives. We observe that both formulations converge at approximately the same rate for high numbers of nodes. However, the precision obtained is not acceptable when considering the computational cost. The precision for the diffuse derivative is still significantly worse.

Figure 29 gives the results for the same domain as that used for the patch test. However, the equation solved and the boundary conditions are chosen in order to give the exact solution $u = sin(2 * pi * x)$. The full derivative version performs reasonably well, while the diffuse derivative diverges.

But the proposed discretization scheme satisfies the patch test at convergence. Both the number of numerical integration points and number of nodes increase. The full derivative must be used. Similar results are obtained when solving an arbitrary problem. The drawback of the method is that the patch test is not verified exactly

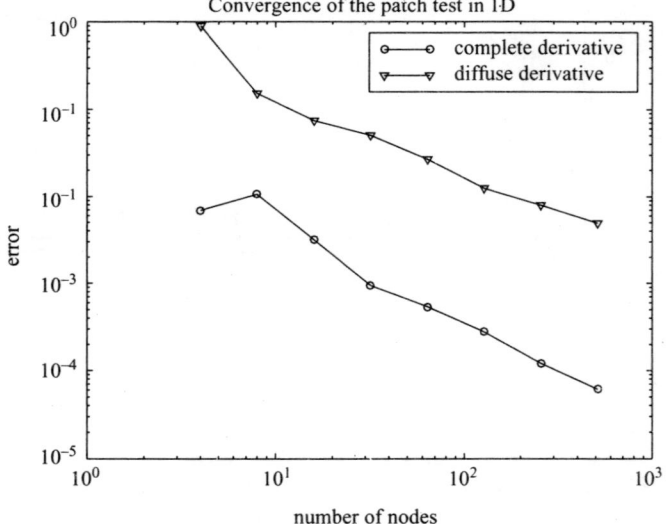

Figure 28 *Patch test*

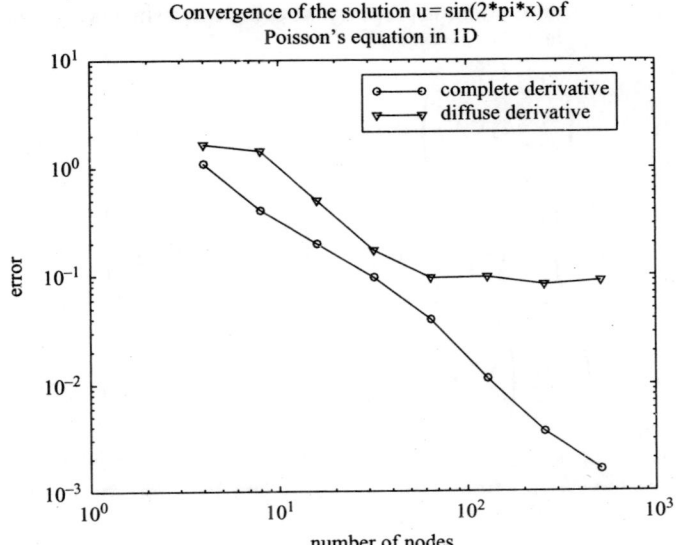

Figure 29 *Solution of Poisson equation*
Convergence in 1D with varying number of nodes, comparison of the full and diffuse derivatives

for low number of nodes and for low number of integration points. This can be explained by the fact that the Gauss Legendre method integrates poorly the shape functions. Therefore, we propose in the following section a custom quadrature scheme for MLS shape functions in order to ensure the properties needed for exact verification of the patch test.

6.2. *Integration constraint*

We focus now on the necessary condition that the formulation has to fulfill in order to satisfy the patch test in the three following cases:

$$u_{ex}(\mathbf{X}) = 1$$
$$u_{ex}(\mathbf{X}) = x \qquad (90)$$
$$u_{ex}(\mathbf{X}) = y$$

We take

$$u_{ap}(x) = u_1 N_1(x) + \cdots + u_n N_n(x) \qquad (91)$$

and we suppose that the global system (88) has a unique solution. In this case we should obtain $u_i = u_{ex}(x_i)$, due to the linear consistency properties of the MLS approximation. The interpolation property is not necessary in this case. Substituting (91) into (87), we get

$$\int_{\Omega} \nabla N_i \left(\sum_j u_j \nabla N_j \right) d\Omega$$

$$- \int_{\Gamma_D} \left\{ N_i \left(\sum_j u_j \frac{\partial N_j}{\partial n} \right) + \left(\sum_j u_j N_j \right) \frac{\partial N_i}{\partial n} \right\} d\Gamma = F_i \qquad (92)$$

with

$$F_i = \int_{\Gamma_N} N_i \left(\sum_j u_j \frac{\partial N_j}{\partial n} \right) d\Gamma - \int_{\Gamma_D} \left(\sum_j u_j N_j \right) \frac{\partial N_i}{\partial n} d\Gamma \qquad (93)$$

When substituting $u = (1 \quad \cdots \quad 1)^T$ and taking into account the consistency properties of the N_i functions (guaranteed by construction) the first of the three conditions (90) is automatically verified.

For a linear field $u_{ex}(x) = x$ we obtain on the RHS of (87)

$$F_i = \int_{\Gamma_N} n_x N_i d\Gamma - \int_{\Gamma_D} x \frac{\partial N_i}{\partial n} d\Gamma \qquad (94)$$

and substituting the further consistency conditions

$$\sum_i x_i N_i = x, \quad \sum_i x_i \frac{\partial N}{\partial x} = 1, \quad \sum_i x_i \frac{\partial N}{\partial y} = 0 \qquad (95)$$

The condition (94) reduces to

$$\int_{\Omega} \frac{\partial N_i}{\partial x} d\Omega = \int_{\Gamma} n_x N_i d\Gamma \qquad (96)$$

that must be satisfied for a linear field $u_{ex}(x) = x$. An analogous analysis for $u_{ex}(x) = y$ gives

$$\int_{\Omega} \frac{\partial N_i}{\partial y} d\Omega = \int_{\Gamma} n_y N_i d\Gamma \qquad (97)$$

and subsequently, for an arbitrary linear field $u_{ex}(x) = ax + by + c$

$$\int_{\Omega} \nabla N_i d\Omega = \int_{\Gamma} N_i \mathbf{n} d\Gamma \qquad (98)$$

has to hold. This can be seen as the expression of the Green-Riemann theorem for the discrete integration. In the numerical program, the integrals are calculated in an approximate way; the precision of the patch test depends on the choice of the numerical integration method. We note also that the interpolation property of the MLS approximation is not necessary for the patch test.

6.3. *Custom integration scheme*

As shown in the explicit expressions (paragraph 3.3), MLS shape functions are not polynomial. Thus, in contrast with the finite element context, the standard Gauss Legendre integration scheme is not suited for the meshfree methods. In the, following, we take into account that the domain is split into a set of integration subdomains called tiles or "diffuse elements" (paragraph 5). We propose below a specific integration scheme denoted by $\int_{\tilde{\Omega}}$ which satisfies the global conditions

$$\int_{\tilde{\Omega}} \frac{\partial}{\partial x} \{N_i(\mathbf{x})\} d\Omega = \int_{\Gamma} N_i(\mathbf{x}(s)) dy(s)$$

$$\int_{\tilde{\Omega}} \frac{\partial}{\partial y} \{N_i(\mathbf{x})\} d\Omega = -\int_{\Gamma} N_i(\mathbf{x}(s)) dx(s) \tag{99}$$

on the "tile by tile" basis. The integration cells Ω_e are chosen in such a way that **interior**$(\Omega_i) \cap$ **interior**$(\Omega_j) = \emptyset$, $i \neq j$ and $\cup_e \Omega_e = \Omega$. In the numerical procedure, the boundary integrals $\int_{\Gamma_e} (\cdot) d\Omega$ are computed using a standard Gauss integration. The specific integrals over Ω_e are noted by $\int \int_{\Omega_e} d\Omega \int^{\sim} (\cdot)$ and are defined so that

$$\int_{\tilde{\Omega}_e} \frac{\partial N_i}{\partial x} d\Omega = \int_{\Gamma_e} N_i dy, \quad \int_{\tilde{\Omega}_e} \frac{\partial N_i}{\partial y} d\Omega = -\int_{\Gamma_e} N_i dx \tag{100}$$

is satisfied over each individual subdomain of integration and for any node i connected to Ω_e. In the summation procedure, the boundary integrals between individual subdomains mutually cancel. Therefore, only non-zero contribution comes from the external boundary and the global condition (99) is satisfied.

The discrete LHS integrals are written as

$$\sum_g \omega_g \frac{\partial N_i}{\partial x} (\mathbf{x}_g) = \int_{\Gamma_e} N_i(\mathbf{x}) dy$$

$$\sum_g \omega_g \frac{\partial N_i}{\partial y} (\mathbf{x}_g) = -\int_{\Gamma_e} N_i(\mathbf{x}) dx \tag{101}$$

where $\mathbf{x}_g$ are the usual Gauss-Legendre integration points and ω_g are the custom integration weights. The above expressions can be presented in a matrix form

$$\mathbf{D}\omega = \mathbf{d} \tag{102}$$

Matrix $\mathbf{D}$ and the column vector $\mathbf{d}$ are obtained directly from (25). The integration scheme is extended in order to integrate polynomials. For this reason, we choose a set of monomials $\mathbf{m} = \{1 \ \ x \ \ y \ \ x^2 \ \ xy \ \ y^2 \ \ \cdots \}$ which have to be exactly integrated

$$\sum_g \omega_g m(\mathbf{x}_g) = \int_{\Omega_e} m(\mathbf{x}) d\Omega \tag{103}$$

which may be written in the matrix form

$$\mathbf{G}\omega = \mathbf{g} \tag{104}$$

We note that the solution of the coupled system (102), (104) is not straightforward due to a poor conditioning. Our experience shows that a practical way consists in minimizing $\|\mathbf{G}\omega - \mathbf{g}\|^2$ under the constraint (102). If the dimension of $\mathbf{m}$ is properly chosen, then the minimum is zero. In this case, the complete system ((102),(104)) is satisfied.

6.4. *Numerical verification of the patch test*

In this section, we present patch test results for the example problems (Figure 30, Figure 31 and Figure 32) given in (Lu *et al.*, 1994) and for an arbitrary domain.

The performances of classical and custom integration schemes are compared using the above problems alternatively with interpolating and non-interpolating shape functions and with the full and diffuse derivatives. The results obtained are given in Tables 1 and 2.

As shown in Table 1 and Table 2, in the case of rectangular domain with regular node pattern, the patch test is always satisfied (within the limits of numerical precision); This is true independently of the choice of other parameters and can be easily explained by symmetry reasons. This property is no longer valid in the case of non rectangular domains or for an irregular distribution of nodes.

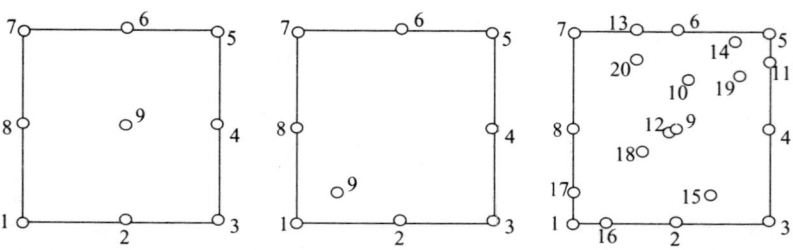

| Figure 30 *Regular grid of nodes* | Figure 31 *Irregular node 9* | Figure 32 *Irregular grid* |

Table 1 *Patch test results using L_2 norm with standard Gauss-Legendre integration; In Figure 31^1 node 9 is located at (0.3, 04), in Figure 31^2 node 9 is located at (0.9, 0.9)*

Gauss weights		Full derivative	Diffuse derivative
MLS approximation	Figure 30	7.7716e-16	8.8818e-16
	Figure 31^1	5:2826e-3	4.3726e-1
	Figure 31^2	1.5014e-4	1.0933e-1
	Figure 32	5.8584e-2	6.7871e + 1
	arbitrary patch	4.0122e-6	3.8446e-3
MLS interpolation	Figure 30	7.7716e-16	0
	Figure 31^1	6.3801e-4	1.2775e-1
	Figure 31^2	4.7010e-4	3.0384e-3
	Figure 32	2.2963e-3	3.4729e-1
	arbitrary patch	1.0930e-5	1.5457e-3

Table 2 *Patch test results using L_2 norm with the custom integration scheme. In Figure 31^1 node 9 is located at (0.3, 04), in Figure 31^2 node 9 is located at (0.9, 0.9)*

Modified weights		Full derivative	Diffuse derivative
MLS approximation	Figure 30	5.5511e-16	5.0653e-11
	Figure 31^1	2.5434e-14	5.9145e-14
	Figure 31^2	2.1663e-16	3.8993e-15
	Figure 32	1.9350e-17	4.3859e-11
	arbitrary patch	9.6859e-14	2.9635e-15
MLS interpolation	Figure 30	3.9968e-15	4.6629e-14
	Figure 31^1	2.0186e-16	2.3012e-14
	Figure 31^2	4.9825e-15	1.8413e-14
	Figure 32	6.4393e-15	6.8530e-10
	arbitrary patch	3.0192e-14	3.4437e-13

The following conclusions can be drawn:

- the modified weights pass the patch test "exactly", for both full and diffuse derivative and for both approximating and interpolating MLS; these results are more accurate than (Dolbow and Belytschko, 1999) obtained in the scope of EFG;
- for the standard Gauss integration, the full derivative is mandatory and the precision of the patch test depends on the density of the Gauss points; the use of the interpolating shape functions improves slightly the results in this case.

However, the computational cost of modified weights is high, as the system (102), (103) has to be solved on each integration domain.

7. Closing remarks

Throughout this paper, we have developed the basis of moving least squares mesh-free approximation and interpolation methods. As an illustration, we have described the diffuse element method for solving PDEs. The meshfree methods do not require an explicit mesh. Only a set of data points and a description of the boundary surfaces are needed. At each evaluation point, a list of nearest nodes is used to approximate the value at that point. The finite element shape functions are replaced by their moving least squares equivalents. This approximation procedure is used to obtain a global system of linear equations. The goal is to achieve a better control of the continuity of the solution, an easier handling of evolving boundaries, the possibilities of adding or removing nodes and the treatment of distorted domains without remeshing.

Despite the undeniable success in many applications, the meshfree methods are still in an early phase of development. The practical implementation of such methods encounters several problems which do not appear in the finite element method. A number of alternative methods in order to take into account the essential boundary conditions reveal advantages and shortcomings. The numerical integration issues must also be clarified. The standard finite element patch test is not a priori satisfied in the case of meshfree methods. The clouds of points used to discretize the domain require an additional treatment in order to establish nodal connectivity. One of the most important theoretical points is the discrete ellipticity of the standard variational formulation. In the extended variational formulation, discrete inf-sup condition has to be established in order to prove convergence. We also have to explain why the patch test integration constraint is crucial in the improvement of numerical results.

The above questions are still open and no definite answers can be given. That is why the topic of meshfree methods gives numerous and exciting opportunities for new ideas and contributions.

8. References

Babuska I., Banerjee U., Osborn J.E., Meshless and Generalized Finite Element Methods: a Survey of Some Major Results, *TICAM Report 02-03, University of Texas at Austin*, January 2002.

Babuska I., Melenk J.M., "The Partition of Unity Method", *International Journal for Numerical Methods in Engineering*, 40, 727–758, 1997.

Barnhill, R.E., *Representation and Approximation of Surfaces in Mathematical Software*, ed. J.R. Rice, Academic Press, New York, 69–120, 1977.

Beissel S., Belytschko T., "Nodal Integration of the Element-Free Galerkin Method", *Computer Methods in Applied Mechanics and Engineering*, 139, 49–74, 1996.

Belytschko T., Krongauz Y., Organ D., Fleming M., and Krysl P., "Meshless Methods: An Overview and Recent Developments", *Computer Methods in Applied Mechanics and Engineering*, 139, 3–47, 1996.

Belytschko T., Lu Y.Y., Gu L., "Element-free Galerkin Methods", *International Journal for Numerical Methods in Engineering*, 37, 229–256, 1994.

Belytschko T., Gu L., Lu. Y.Y., "Fracture and Crack Growth by Element-free Galerkin Methods. Modelling and Simulation", *Material Science and Engineering*, 2, 519–534, 1994(a).

Bonet J., Lok T.-S. L., "Variational and Momentum Preservation Aspects of Smooth Particle Hydrodynamics Formulation", *Computer Methods in Applied Mechanics and Engineering*, 180, 97–115, 1999.

Breitkopf P., Rassineux A., Touzot G., Villon P., "Explicit Form and Efficient Computation of MLS Shape Functions and Their Derivatives", *International Journal for Numerical Methods in Engineering*, 48, 451–456, 2000.

Breitkopf P., Rassineux A., Villon P., Saannouni K., Cherouat H., "Meshfree Operators for Consistent Field Transfer in Large Deformation Plasticity", *ECCOMAS-ECCM-2001*, Cracow, Poland, 26–29 June 2001.

Breitkopf P., Touzot G., Villon P., "Double Grid Diffuse Collocation Method", *Computational Mechanics*, 25, No 2/3, 199–206, 2000.

Chen J-S, Han W., You Y., Meng X., "A Reproducing Kernel Method with Nodal Interpolation Property", *International Journal for Numerical Methods in Engineering*, in press, 2002.

Chen J.S., Wu C.T., Yoon S., You Y., "Nonlinear Version of Stabilized Conforming Nodal Integration for Galerkin Meshfree Methods", *International Journal for Numerical Methods in Engineering*, 53, 2587–2615, 2002.

Cleveland, W.S., "Robust Locally Weighted Regression and Smoothing Scatterplots", *Journal of the American Statistical Association*, December, Vol, 74, No. 368, 829–836, 1979.

De S., Bathe K.J., "The Method of Finite Spheres", *Computational Mechanics*, 25, 329–345, 2000.

Dolbow J., Belytschko T., "Numerical Integration of the Galerkin Weak Form in Meshfree methods", *Computational Mechanics*, 23(3), 219–230, 1999.

Gordon William J., Wixom James A., "Shepard's Method of Metric Interpolation to Bivariate and Multivariate Interpolation", *Math. Comp.* 32(141), 253–264, 1978.

Huerta A., Vidal Y., Villon P., "Locking in the Incompressible Limit: Pseudo-Divergence-Free Element-free Galerkin", *Proceedings of the Fifth World Congress on Computational Mechanics (WCCM V)*, July 7–12, 2002, Vienna, Austria.

Krige, D.G., "Two-dimensional Weighted Moving Average Trend Surfaces for Ore Evaluation", *Journal of the South African Institute of Mining & Metallurgy*, 67, 13–79, 1966.

Lancaster P., Salkauskas K., *Curve and Surface Fitting: an Introduction*, Academic Press, London, Orlando, 1986.

Lancaster P., Salkauskas K., "Surfaces Generated by Moving Least Squares Methods", *Math. Comp.* 37, 141–158, 1981.

Lin H., Atluri S.N., "Meshless Local Petrov-Galerkin (MLPG) Method for Convection – Diffusion Problems", *Computer Modeling in Engineering & Sciences*, 1(2), 45–60, 2000.

Liszka T., Orkisz J., "The Finite Difference Method at Arbitrary Irregular Grids and its Application in Applied Mechanics", *Computers and Structures*, 11, 83–95, 1980.

Liu W.K., Chen Y., Jun S., Chen J.S., Belytschko T., Pan C., Uras R.A. Chang C.T., "Overview and Applications of the Reproducing Kernel Particle Methods", *Archives of Computational Methods in Engineering: State of the art reviews*, 3, 3–80, 1996.

Lu Y.Y., Belytschko T., Gu L., "A New Implementation of the Element Free Galerkin Method", *Computer Methods in Applied Mechanics and Engineering*, 113, 397–414, 1994.

Lucy L.B., "A Numerical Approach to the Testing of the Fission Hypothesis", *Astronomical Journal* 82, 1013–1024, 1977.

Mac Lain D.H., "Drawing Contours with Arbitrary Data Points", *The Computer Journal*, 17(4), 318–324, 1974.

Matheron G., "Principles of Geostatistics", *Economic Geology*, 58, 1246–1266, 1963.

Mukherjee Y.X., Mukherjee S., "On Boundary Conditions in Element-free Galerkin Method", *Computational Mechanics*, 11, 1997, 264–270.

Nayroles B., Touzot G., Villon P., "Generalizing the Finite Element Method: Diffuse Approximation and Diffuse Elements", *Computational Mechanics*, 10, 307–318, 1992.

Oñate E., Idelsohn S.R., "A Mesh-free Finite Point Method for Advective-diffusive Transport and fluid flow problems", *Computational Mechanics*, 21, 283–292, 1998.

Savignat J-M, *Approximation Diffuse Hermite et ses Applications*, Thèse de Doctorat, Ecole des Mines de Paris, October 2000.

Shepard D., "A two-dimensional Interpolation Function for Irregularly Spaced Data", *Proc. 23rd National Conference ACM*, 517–524, 1968.

Sukumar N., Moës N., Moran B., Belytschko T., "Extended Finite Element Method for Three-Dimensional Crack Modeling", *International Journal for Numerical Methods in Engineering*, 48(11), 1549–1570, 2000.

Sulsky D., Schreyer H.L., The Particle-In-Cell Method as a Natural Impact Algorithm, Sandia National Laboratories, Contract No. AC-1801, 1993.

Syczewski M., Tribillo R., "Singularities of Sets Used in the Mesh Method", *Computers and Structures*, 14(5–6), 509–511, 1981.

Villon P, *Contribution à l'Optimisation*, Thèse de Docteur d'Etat, Université de Technologie de Compiègne, France, 1991.

Wyatt M.J., Davies G., Snell C., "A New Difference Based Finite Element Method", *Instn. Engineers*, 59(2), 395–409, 1975.

Zhang X., Liu X-H., Song K-Z., Lu M-W., Least-squares Collocation Meshless Method, *International Journal for Numerical Methods in Engineering*, 51, 1089–1100, 2001.

Chapter 2

Locking in the Incompressible Limit: Pseudo-divergence-free Element-free Galerkin

Yolanda Vidal & Antonio Huerta
Departament de Matemàtica Aplicada III, Laboratori de Càlcul Numeric (LaCàN), Universitat Politècnica de Catalunya, Barcelona, Spain

Pierre Villon
Laboratoire de Mécanique Roberval, Université de Technologie de Compiègne, Compiègne, France

1. Introduction

Finite element methods require a mesh with a minimum quality. Reasonable meshes may be difficult in certain contexts, for instance, problems involving large deformations, crack propagation, discontinuities and adaptive processes. Recently, meshless methods have been developed to overcome these difficulties; among them are the *smooth particle hydrodynamics* (SPH) (Lucy, 1977; Monaghan, 1988; Randles and Libersky, 1996), *the diffuse element method* (DEM) (Nayroles, Touzot and Villon, 1992), *the element-free Galerkin method* (EFG) (Belytschko, Krongauz, Organ, Fleming and Krysl, 1996; Belytschko, Lu and Gu, 1994; Belytschko and Tabarra, 1996; Lu, Belytschko and Gu, 1994), the *reproducing kernel particle method* (Liu, Chen, Uras and Chang, 1996; Liu, Jun and Zhang, 1995; Liu, Jun, Adee and Belytschko, 1995; Liu, Li and Belytschko, 2000), the *HP cloud method* (Duarte and Oden, 1995; Duarte and Oden, 1996) and the *partition of unity method* (Melenk and Babuska, 1996).

The objective of meshless methods is to suppress some of the overhead due to mesh generation by constructing the approximation entirely in terms of nodes without defining *a priori* connectivities between them. Due to the flexibility in constructing the conforming shape functions to meet specific needs for different applications, it has been reported that the meshless methods are particulary suitable for crack propagation, *hp* adaptivity and large deformation problems.

Locking of standard finite elements has been widely studied. It appears because poor numerical interpolation leads to an overconstrained system. It is acknowledged that in a displacement based finite element method, linear approximations perform poorly for the modeling of incompressible materials. For incompressible, or nearly incompressible, materials an additional constraint appears in the field equations which requires the divergence of the displacement field to be zero in the domain. This constraint is difficult to fulfill for low order elements. Locking is attenuated and can be suppressed for increasing polynomial degrees; in the context of an hp adaptive strategy, Babuska and Suri (1992) and Suri (1996) present a review on this issue. Moreover, several techniques are available to alleviate or completely remove the locking phenomena in finite element approximations (see Hughes, 2000).

However, locking in meshless methods is still an open topic. Even recently, Zhu and Atluri (1998) claimed that meshless methods do not exhibit volumetric locking. Now it is clear that this is not true. For instance, Dolbow and Belytschko (1999) analyze the EFG method using the numerical infsup condition. Moreover, several authors claim that increasing the dilation parameter locking phenomena in mesh-free methods can be suppressed, or at least attenuated. Their argument is based on numerical examples (Askes, de Borst and Heeres, 1999; Dolbow and Belytschko, 1999) or on the heuristic constraint ratio (Chen, Yoon, Wang and Liu, 2000) proposed by Hughes (2000). In a recent paper by Huerta and Fernández Méndez (2001) this issue is clarified, determining the influence of the dilation parameter on the locking behavior of EFG near the incompressible limit. This is done performing a modal analysis: studying the fundamental modes (base of the solution space) and their corresponding energy (eigenvalue). In particular, EFG behavior is compared with standard finite elements, bilinear and biquadratic. It concludes that an increase of the dilation parameter attenuates, but never suppresses, the volumetric locking and that, as in standard finite elements, an increase in the order of reproducibility reduces the relative number of locking modes.

Thus, large domains of influence alleviate locking but for small domains of influence; however, the direct application of the EFG approximation can result in volumetric locking. In dynamic problems and many nonlinear problems, small domains of influence are preferred because they improve the local resolution and enhance the sparsity of the system of equations. Therefore, procedures which avoid locking, even for small domains of influence, are needed. Until now the remedies proposed in the literature are extensions of the methods developed for finite elements.

As noticed before, there are several techniques available to alleviate or remove the locking phenomena in finite element approximations. For example, Suri (1996) shows that locking can be alleviated through the use of higher-order p elements. Alternatively, locking can be removed by mixed methods in which different approximations are implemented for the displacement and pressure fields (see, for instance, Hermann, 1965; Hughes, 2000). However, mixed methods are more expensive due to the need for additional unknowns. Alternatives which do not require additional degrees of freedom are selective reduced integration or strain projection methods. Extensions of these techniques to meshless methods can be found. For example, Dolbow and Belytschko (1999) propose a new formulation of the EFG method using a selective reduced integration and Chen et al (2000) suggest an improved *reproducing kernel particle method* (RKPM) using a pressure projection method.

Here a novel approach is explored. It consists in using interpolation functions that verify approximately the divergence-free constraint. These interpolating functions can be defined *a priori* and are independent of the particle distribution. Moreover, as the density of particles is increased the divergence-free condition is better approximated. This method is based on diffuse derivatives (see Nayroles et al., 1992), which, as proved by Villon (1991), converge to the derivatives of the exact solution when the radius of the support goes to zero (for a fixed dilation parameter).

2. Diffuse derivatives

2.1. *Preliminaries of the EFG method*

This section will not be devoted to developing or discussion of meshfree methods in detail or their relation with moving least squares (MLS) interpolants. There are well known references with excellent presentations of meshfree methods (see, for instance, Belytschko, Krongauz, Organ, Fleming and Krysl, 1996; Liu, Belytschko and Oden, editors, 1996; Liu, Chen, Jun, Chen, Belytschko, Pan, Uras and Chang, 1996; Liu et al., 2000; Nayroles et al., 1992). Here some basic notions will be recalled in order to introduce the notation and the approach employed in following sections.

The moving least squares approach is based on the local (*i.e.*, at any point z in the neighborhood of x) approximation of the unknown scalar function $u(z)$ by u^ρ as

$$u(z) \simeq u^\rho(x, z) = \mathbf{P}^\mathrm{T}(z)\mathbf{a}(x) \quad \text{for } z \text{ near } x \tag{1}$$

where the coefficients $\mathbf{a}(x) = \{a_0(x), a_1(x), \ldots, a_l(x)\}^\mathrm{T}$ are not constant; they depend on point x, and $\mathbf{P}(z) = \{p_0(z), p_1(z), \ldots, p_l(z)\}^\mathrm{T}$ includes a complete basis of the subspace of polynomials of degree m. In one dimension, it is usual that $p_i(x)$ coincides with the monomials x^i, and, in this particular case, $l = m$. The coefficients $\mathbf{a}$ are obtained by minimization of the functional $J_x(\mathbf{a})$ centered in x and defined as

$$J_x(\mathbf{a}) = \sum_{i \in I_x} \phi(x, x_i)[u(x_i) - \mathbf{P}(x_i)\mathbf{a}(x)]^2 \tag{2}$$

where $\phi(x, x_i)$ is a weighting function (positive, even and with compact support) which characterizes the mesh-free method. For instance, if $\phi(x, x_i)$ is continuous together with its first k derivatives, the interpolation is also continuous together with its first k derivatives. The particles cover the computational domain Ω, $\Omega \subset \mathbb{R}^{nsd}$, and, in particular, a number of particles $\{x_i\}_{i \in I_x}$ belong to the support of $\phi(x, x_i)$. The minimization of $J_x(a)$ induces the standard normal equations in a weighted least-squares problem

$$\mathbf{M}(x)\mathbf{a}(x) = \sum_{i \in I_x} \phi(x, x_i)u(x_i)\mathbf{P}(x_i) \tag{3}$$

where, as usual, the Gram matrix $\mathbf{M}(x)$ is the scalar product of the interpolation polynomials:

$$\mathbf{M}(x) = \sum_{i \in I_x} \phi(x, x_i)\mathbf{P}(x_i)^{\mathsf{T}}\mathbf{P}(x_i).$$

That is,

$$\langle u, v \rangle = \sum_{i \in I_x} \phi(x, x_i)u(x_i)v(x_i) \tag{4}$$

must define a discrete scalar product. Thus, several conditions on the particle distribution are implicitly assumed (see, for instance, Huerta and Fernández-Méndez, 2000).

Once the normal equations, Eqs (3), are solved the coefficients $\mathbf{a}$ are substituted in (1). Since the weighting function ϕ usually favors the central point x, it seems reasonable to assume that such an approximation is more accurate precisely at $z = x$ and thus the approximation (1) is particularized at x, that is,

$$u(x) \simeq u^\rho(x) = \mathbf{P}^{\mathsf{T}}(x)\mathbf{a}(x) = \mathbf{P}^{\mathsf{T}}(x)\mathbf{M}^{-1}(x) \sum_{i \in I_x} \phi(x, x_i)u(x_i)\mathbf{P}(x_i). \tag{5}$$

This expression can also be written in a standard interpolation form

$$u^\rho(x) = \sum_{i \in I_x} N_i^\rho(x)u(x_i) = \sum_{i \in I_x} \underbrace{[\phi(x, x_i)\mathbf{P}^{\mathsf{T}}(x)\mathbf{M}^{-1}(x)\mathbf{P}(x_i)]}_{N_i^\rho(x)}u(x_i) \tag{6}$$

2.2. The diffuse derivative

The approximation of the derivative of u is the derivative of u^ρ. This requires one to derive (5), that is

$$\frac{\partial u}{\partial x_i} \simeq \frac{\partial u^\rho}{\partial x_i} = \frac{\partial \mathbf{P}^{\mathsf{T}}}{\partial x_i}\mathbf{a}(x) + \mathbf{P}^{\mathsf{T}}\frac{\partial \mathbf{a}}{\partial x_i} \quad \text{for } i = 1, \dots, n_{sd}. \tag{7}$$

Note that the derivative of the polynomials in $\mathbf{P}$ is trivial but the derivative of the coefficients $\mathbf{a}$ requires the resolution of a linear system of equations with the same matrix $\mathbf{M}$. Moreover, the derivatives of the polynomials can be evaluated *a priori* but the derivatives of the coefficients require the knowledge of the cloud of particles surrounding each point x.

Thus the concept of diffuse derivative proposed by and defined as

$$\frac{\delta u^\rho}{\delta x_i} = \frac{\partial u^\rho}{\partial z_i}\bigg|_{z=x} = \frac{\partial \mathbf{P}^\mathsf{T}}{\partial z_i}\bigg|_{z=x} \mathbf{a}(x) = \frac{\partial \mathbf{P}^\mathsf{T}}{\partial x_i}\mathbf{a}(x) \quad \text{for } i = 1, \dots, n_{sd}$$

is, from a computational cost point of view, an interesting alternative to (7). Moreover, it shows that the diffuse derivative converges at optimal rate to the derivative of u.

Proposition. *If u^ρ is an approximation to u with an order of consistency m (i.e. $\mathbf{P}$ includes a complete basis of the subspace of polynomials of degree m) and ρ/h is constant, then*

$$\left\|\frac{\partial^{|k|} u}{\partial x^k} - \frac{\delta^{|k|} u^\rho}{\delta x^k}\right\|_\infty \le C(x)\frac{\rho^{m+1-|k|}}{(m+1)!} \quad \forall |k| = 0, \dots, m. \tag{8}$$

where k is a multi-index, $k = (k_1, k_2, \dots, k_{n_{sd}})$ and $|k| = k_1 + k_2 + \dots + k_{n_{sd}}$.

Proof. Lets assume $u \in C^{m+1}(\overline{\Omega})$ where C^{m+1} is the space of $(m+1)$ times continuously differentiable functions. Recall that Taylor's formula of order m can be written as:

$$u(x + h) = \sum_{|\alpha|=0}^{m} \frac{1}{\alpha!} h^\alpha \frac{\partial^{|\alpha|} u}{\partial x^\alpha}(x) + R_{m+1}(x + \theta h), \tag{9}$$

where $\theta \in]0, 1[$, $R_{m+1}(x + \theta h)$ is the error term and α is a multi-index such that,

$$h^\alpha := h_1^{\alpha_1} h_2^{\alpha_2} \cdots h_{n_{sd}}^{\alpha_{n_{sd}}}; \quad \alpha! := \alpha_1! \alpha_2! \cdots \alpha_{n_{sd}}!; \quad |\alpha| = \alpha_1 + \alpha_2 + \dots + \alpha_{n_{sd}}$$

Equation (9) can be rewritten taking $z = x + h$

$$u(z) = \sum_{|\alpha|=0}^{m} \frac{1}{\alpha!} \left(\frac{z-x}{\rho}\right)^\alpha \rho^\alpha \frac{\partial^{|\alpha|} u}{\partial x^\alpha}(x) + R_{m+1}(x, z).$$

Thus, Taylor's formula can also be written as:

$$u(z) = \mathbf{P}^\mathsf{T}\left(\frac{z-x}{\rho}\right)\mathbf{U}(x) + R_{m+1}(x, z), \tag{10}$$

where

$$\mathbf{P}(\xi) = \left\{ \frac{\xi^\alpha}{\alpha!} \right\}; \quad \mathbf{U}(x) = \left\{ \rho^\alpha \frac{\partial^{|\alpha|} u}{\partial x^\alpha} \right\} \quad |\alpha| = 0, \ldots, m. \tag{11}$$

Observe that $\mathbf{U}(x)$ depends on the exact derivatives of u.

The MLS approach is based on the local approximation of the unknown scalar function u by u^p, see equation (1). Since in equation (10) polynomials $\mathbf{P}(\xi)$ are centered and scaled, the MLS interpolant is also centered and scaled,

$$u(z) \simeq u^p(x, z) = \mathbf{P}^\mathrm{T} \left(\frac{z - x}{\rho} \right) \mathbf{a}(x) \quad \text{for } z \text{ near } x.$$

Then the MLS approach requires the resolution of the normal equations given by (3), here $u(x_i)$ is substituted by (10)

$$\mathbf{M}(x)\mathbf{a}(x) = \left\langle \mathbf{P} \left(\frac{z - x}{\rho} \right), \mathbf{P}^\mathrm{T} \left(\frac{z - x}{\rho} \right) \mathbf{U}(x) + R_{m+1}(x, z) \right\rangle,$$

which can be rearranged as

$$\mathbf{M}(x)[\mathbf{a}(x) - \mathbf{U}(x)] = \sum_{j \in I_x} \phi \left(\frac{x_j - x}{\rho} \right) \mathbf{P} \left(\frac{x_j - x}{\rho} \right) R_{m+1}(x, x_j) =: \mathbf{b}. \tag{12}$$

Now, let us rewrite the r.h.s. of (12) in a more convenient way. The error term of the Taylor's formula has the form

$$R_{m+1}(x, x_j) = \sum_{|\alpha|=m+1} \frac{(x_j - x)^\alpha}{(m + 1)!} \frac{\partial^{|\alpha|} u}{\partial x^\alpha}(x, x_j), \tag{13}$$

substituting (13) in the definition of vector $\mathbf{b}$, see (12), produces

$$\mathbf{b} = \sum_{j \in I_x} \phi \left(\frac{x_j - x}{\rho} \right) \mathbf{P} \left(\frac{x_j - x}{\rho} \right) \sum_{|\alpha|=m+1} \frac{(x_j - x)^\alpha}{(m + 1)!} \frac{\partial^{|\alpha|} u}{\partial x^\alpha}(x, x_j).$$

Each component of the previously defined vector $\mathbf{b}$ is associated to the corresponding component of $\mathbf{P}$, namely the polynomial of degree $|k| = 0, \ldots, m$ defined as

$$\xi^k / k! = (\xi_1^{k_1} \xi_2^{k_2} \cdots \xi_{n_{sd}}^{k_{n_{sd}}}) / (k_1! k_2! \cdots k_{n_{sd}}!).$$

Under these circumstances, each component of **b** can be written as

$$b_k = \sum_{j \in I_x} \phi\left(\frac{x_j - x}{\rho}\right) \frac{(x_j - x)^k}{\rho^{|k|}} \frac{1}{|k|!} \sum_{|\alpha|=m+1} \frac{(x_j - x)^\alpha}{(m+1)!} \frac{\partial^{|\alpha|} u}{\partial x^\alpha}(x, x_j)$$

$$= \frac{\rho^{m+1}}{(m+1)!} \frac{1}{|k|!} \underbrace{\sum_{j \in I_x} \phi\left(\frac{x_j - x}{\rho}\right) \sum_{|\alpha|=m+1} \left(\frac{x_j - x}{\rho}\right)^{k+\alpha} \frac{\partial^{|\alpha|} u}{\partial x^\alpha}(x, x_j)}_{r_k(x)}$$

$$= \frac{\rho^{m+1}}{(m+1)!} r_k(x). \tag{14}$$

Thus, the r.h.s. of (12) becomes

$$\mathbf{b} = \frac{\rho^{m+1}}{(m+1)!} \mathbf{r}(x). \tag{15}$$

Substituting (15) into equation (12) and assuming that **M** is regular,

$$\mathbf{a}(x) - \mathbf{U}(x) = \frac{\rho^{m+1}}{(m+1)!} \mathbf{M}^{-1}(x)\mathbf{r}(x).$$

On one hand, r_k is bounded for all $|k| = 0, \ldots, m$. This can be seen from the definition of r_k, see (14). Note that for a fixed x, if ρ/h is constant, r_k is the sum of products of continuous functions in Ω. Thus, it is a continuous function in Ω. Moreover, in every product, there is the weighting function ϕ, which has compact support. Since r_k is a continuous function in a compact support it is bounded by a constant that only depends on x.

On the other hand, matrix **M** is also bounded (see Huerta and Fernández-Méndez, 2000). Then, if both **M** and r_k, are bounded, a constant $C(x)$ can be defined as the bound of $\mathbf{M}^{-1}(x)\mathbf{r}(x)$ and consequently

$$|\mathbf{a}(x) - \mathbf{U}(x)| \leq \frac{\rho^{m+1}}{(m+1)!} C(x).$$

The previous expression can be divided by $\rho^{|k|}$. Then, for each component,

$$\left| \frac{a_k(x)}{\rho^{|k|}} - \frac{U_k(x)}{\rho^{|k|}} \right| \leq \frac{\rho^{m+1-|k|}}{(m+1)!} C(x) \quad \forall |k| = 0, \ldots, m, \tag{16}$$

where a_k and U_k are the components of **a** and **U**, respectively. Recall that each component of $\mathbf{U}(x)$ depends on the corresponding exact derivatives of u, see (11).

Now, observe that each component of $\mathbf{a}(x)$ shall depend on the corresponding pseudo-derivatives; that is, for $|k| = 0, \ldots, m$

$$\frac{\delta^{|k|}u^\rho}{\delta x^k} := \frac{\delta^{|k|}u^\rho}{\delta x_1^{k_1} \cdots \delta x_{n_{sd}}^{k_{n_{sd}}}} := \frac{\partial^{|k|}u^\rho}{\partial z_1^{k_1} \cdots \partial z_{n_{sd}}^{k_{n_{sd}}}}\bigg|_{z=x} = \frac{\mathbf{a}_k(x)}{\rho^{k_1} \cdots \rho^{k_{n_{sd}}}}. \qquad (17)$$

Finally, replacing the definition of $\mathbf{U}(x)$ and $\mathbf{a}(x)$ given by (11) and (17), one gets the final expression, which completes the proof,

$$\left\| \frac{\partial^{|k|}u}{\partial x^k} - \frac{\delta^{|k|}u^\rho}{\delta x^k} \right\|_\infty \leq C(x)\frac{\rho^{m+1-|k|}}{(m+1)!} \qquad \forall |k| = 0, \ldots, m. \qquad \square$$

3. Pseudo-divergence free condition

3.1. *Diffuse divergence*

In the previous section the diffuse derivative was introduced and its convergence to the actual derivative as $\rho \to 0$ was proved. Incompressible computations require that the approximating field must be divergence free. That is, the solution $\mathbf{u}(x)$, now a vector $\mathbf{u} : \mathbb{R}^{n_{sd}} \to \mathbb{R}^{n_{sd}}$, verifies $\nabla \cdot \mathbf{u} = 0$, and the approximation $\mathbf{u}^\rho(x)$ should also be divergence-free. This condition however depends on the interpolation space. Here, instead of requiring a divergence-free interpolation, the diffuse divergence of the approximation

$$\mathbf{u}^\rho = \begin{pmatrix} u_1^\rho \\ \vdots \\ u_{n_{sd}}^\rho \end{pmatrix} = \begin{pmatrix} \mathbf{P}^T\mathbf{a}_1 \\ \vdots \\ \mathbf{P}^T\mathbf{a}_{n_{sd}} \end{pmatrix}$$

$$= \begin{pmatrix} p_0(x)\mathbf{I}_{n_{sd}} & p_1(x)\mathbf{I}_{n_{sd}} & \cdots & p_l(x)\mathbf{I}_{n_{sd}} \end{pmatrix} \begin{pmatrix} \mathbf{c}_0(x) \\ \mathbf{c}_1(x) \\ \vdots \\ \mathbf{c}_l(x) \end{pmatrix}$$

$$= \mathbf{Q}^T\mathbf{c}$$

is imposed equal to zero, that is

$$\nabla^\delta \cdot \mathbf{u}^\rho = \sum_{i=1}^{n_{sd}} \frac{\delta \mathbf{u}^\rho}{\delta x_i} = \sum_{i=1}^{n_{sd}} \frac{\partial \mathbf{P}^T}{\partial x_i}\mathbf{a}_i(x) = (\nabla \cdot \mathbf{Q}^T(x))\mathbf{c}(x) = 0. \qquad (18)$$

Note that $\mathbf{I}_{n_{sd}}$ is the identity matrix of order n_{sd} and the coefficients have been rearranged as

$$\mathbf{c}^T = (\underbrace{a_{0,1} \cdots a_{0,n_{sd}}}_{\mathbf{c}_0^T(x)} \quad \underbrace{a_{1,1} \cdots a_{1,n_{sd}}}_{\mathbf{c}_1^T(x)} \quad \cdots \quad \underbrace{a_{l,1} \cdots a_{l,n_{sd}}}_{\mathbf{c}_l^T(x)}).$$

Equation (18) must hold at each point x and for any approximation. Thus appropriate interpolation functions, $\mathbf{Q}$, must be defined in order to verify (18) and thus ensure asymptotically a divergence-free interpolation (*i.e.*, the divergence-free condition is fulfilled as $\rho \to 0$).

3.2. *A 2D pseudo-divergence free interpolation*

The previous concepts are particularized for a two-dimensional case and in order to define pseudo-divergence-free interpolation functions. Suppose for instance that consistency of order two is desired, then $\mathbf{P}^T = \{1, x_1, x_2, x_1^2, x_1 x_2, x_2^2\}$, thus

$$\mathbf{Q}^T = \begin{pmatrix} 1 & 0 & x_1 & 0 & x_2 & 0 & x_1^2/2 & 0 & x_1 x_2 & 0 & x_2^2/2 & 0 \\ 0 & 1 & 0 & x_1 & 0 & x_2 & 0 & x_1^2/2 & 0 & x_1 x_2 & 0 & x_2^2/2 \end{pmatrix} \tag{19}$$

and

$$\mathbf{c}^T = \begin{pmatrix} a_{0,1} & a_{0,2} & a_{1,1} & a_{1,2} & a_{2,1} & a_{2,2} & a_{3,1} & a_{3,2} & a_{4,1} & a_{4,2} & a_{5,1} & a_{5,2} \end{pmatrix}. \tag{20}$$

The pseudo-divergence-free condition (18) is, in this case, written as

$$\nabla^\delta \cdot u^\rho = \frac{\partial \mathbf{P}^T}{\partial x_1} \mathbf{a}_1 + \frac{\partial \mathbf{P}^T}{\partial x_2} \mathbf{a}_2 = 0,$$

which implies

$$(a_{1,1} + a_{2,2}) + x_1(a_{3,1} + a_{4,2}) + x_2(a_{4,1} + a_{5,2}) = 0,$$

and consequently

$$a_{1,1} + a_{2,2} = 0, \quad a_{3,1} + a_{4,2} = 0, \quad a_{4,1} + a_{5,2} = 0.$$

The influence of these three restrictions in the interpolation functions (19) can be viewed as follows

$$\begin{pmatrix} 1 & 0 & x_1 & 0 & x_2 & 0 & x_1^2/2 & 0 & x_1 x_2 & 0 & x_2^2/2 & 0 \\ 0 & 1 & -x_2 & x_1 & 0 & 0 & -x_1 x_2 & x_1^2/2 & -x_2^2/2 & 0 & 0 & 0 \end{pmatrix}, \tag{21}$$

where one should note that the coefficients in the x_1 and x_2 directions are now coupled and that the total number of degrees of freedom has decreased.

3.3. *The pseudo-divergence free EFG method*

Using (21), let $\mathbf{Q}_\delta$ be the new interpolation matrix (where obviously the unnecessary columns have been removed). The interpolation is then defined as

$$u(z) \simeq u^\rho(x,z) = \begin{pmatrix} u_1^\rho(x,z) \\ u_2^\rho(x,z) \end{pmatrix} = \mathbf{Q}_\delta^{\mathrm{T}}(z)\mathbf{c}(x). \tag{22}$$

The vector version of the discrete scalar product defined in (4),

$$\langle u, v \rangle = \sum_{i \in I_x} \phi(x, x_i) u^{\mathrm{T}}(x_i) v(x_i)$$

allows us now to reproduce the MLS approximation. Thus at each point x the normal equations should be solved, see (3),

$$\mathbf{M}(x)\mathbf{c}(x) = \langle u, \mathbf{Q}_\delta \rangle \quad \text{with} \quad \mathbf{M}(x) := \langle \mathbf{Q}_\delta, \mathbf{Q}_\delta \rangle.$$

Thus, as previously, the coefficients $\mathbf{c}$ are substituted in (22) and the approximation is particularized at $z = x$. Then, equation (5) becomes

$$u(x) \simeq u^\rho(x) = \mathbf{Q}_\delta^{\mathrm{T}}(x)\mathbf{c}(x) = \mathbf{Q}_\delta^{\mathrm{T}}(x)\mathbf{M}^{-1}(x)\langle u, \mathbf{Q}_\delta \rangle,$$

and a final expression similar to (6) can be found as

$$u^\rho(x) = \sum_{i \in I_x} \mathbf{N}_i^\rho(x)u(x_i) = \sum_{i \in I_x} \left[\phi(x, x_i)\mathbf{Q}_\delta^{\mathrm{T}}(x)\mathbf{M}^{-1}(x)\mathbf{Q}_\delta(x_i) \right] u(x_i).$$

It is important to note that the matrix of interpolation functions $\mathbf{N}_i^\rho$ is now a full matrix not a diagonal one as standard EFG would induce in this non scalar problem. This is due to the fact that the two components of the solution are linked by the incompressibility restriction.

4. Modal analysis

4.1. *Preliminaries*

The modal analysis presented here follows the same rationale originally presented by Huerta and Fernández-Méndez (2001). It is restricted to small deformations, namely $\nabla^s u$, where u is the displacement and ∇^s the symmetric gradient, *i.e.* $\nabla^s = \frac{1}{2}(\nabla^{\mathrm{T}} + \nabla)$. Moreover, linear elastic isotropic materials under plane strain conditions are considered. Dirichlet boundary conditions are imposed on Γ_D, a traction h is prescribed along the Neumann boundary Γ_N and there is a body force f. Thus, the problem that needs to be solved may be stated as: solve for $u \in [H^1_{\Gamma_D}]^2$ such that

$$\frac{E}{1+\nu} \int_\Omega \nabla^s v : \nabla^s u \, d\Omega + \frac{E\nu}{(1+\nu)(1-2\nu)} \int_\Omega (\nabla \cdot v)(\nabla \cdot u) \, d\Omega$$

$$= \int_\Omega f \cdot v \, d\Omega + \int_{\Gamma_N} h \cdot v \, d\Gamma \quad \forall v \in [H^1_{0,\Gamma_D}]^2. \quad (23)$$

In this equation, the standard vector subspaces of H^1 are employed for the solution u

$$[H^1_{\Gamma_D}]^2 := \{u \in [H^1]^2 | u = u_D \text{ on } \Gamma_D\}$$

(Dirichlet conditions, u_D, are automatically satisfied) and for the test functions v

$$[H^1_{0,\Gamma_D}]^2 := \{v \in [H^1]^2 | v = 0 \text{ on } \Gamma_D\}$$

(zero values are imposed along Γ_D).

This equation shows the inherent difficulties of the incompressible limit. The standard *a priori* error estimate emanating from (23) and based on the energy norm, which is induced by the LHS of (23), is

$$\| u - u_h \| \le \inf_{w \in S_h} \| u - w \| \le C_{u,\nu,p} h^{f(p)} \quad (24)$$

where S_h is the finite dimensional subspace of $[H^1_{\Gamma_D}]^2$ in which the approximation u_h is sought, $C_{u,\nu,p}$ is a constant independent of h (characteristic size of the mesh), and $f(p)$ is a positive monotone function of p (degree of the polynomials used for the interpolation). The subindices of the constant C indicate that it depends on the Poisson ratio, the order of the interpolation and the exact solution itself.

From (23) one can observe the difficulties of the energy norm to produce a small infimum in (24) for values of ν close to 0.5. In fact, in order to have finite values of the energy norm the divergence-free condition must be enforced in the continuum case, *i.e.* $\nabla \cdot u = 0$ for $u \in [H^1_{\Gamma_D}]^2$, and also in the finite dimensional space, *i.e.* $\nabla \cdot u_h = 0$ for $u_h \in S_h \subset [H^1_{\Gamma_D}]^2$. In fact, locking will occur when the approximation space S_h is not rich enough for the approximation to verify the divergence-free condition.

Under these conditions, it is evident that locking may be studied from the LHS of (23). This is the basis for the modal analysis of locking. The discrete eigenfunctions (the eigenvectors) corresponding to the LHS of (23) are computed because they completely describe, in the corresponding space, the behavior of the bilinear operator induced by this LHS.

In computational mechanics it is standard to write the strain, $\boldsymbol{\varepsilon}$, and the stress, $\boldsymbol{\sigma}$, tensors in vector form (Belytschko, Liu and Moran, 2000). Moreover, under the assumptions already discussed, they are related as

$$\boldsymbol{\varepsilon} = \mathbf{Bd}, \quad \boldsymbol{\sigma} = \mathbf{C}\boldsymbol{\varepsilon}, \quad \mathbf{C} = \frac{E}{(1+v)(1-2v)} \begin{pmatrix} 1-v & v & 0 \\ v & 1-v & 0 \\ 0 & 0 & (1-2v)/2 \end{pmatrix}.$$

Where $\mathbf{d}$ is the vector of nodal displacements (the coefficients corresponding to the approximation u_h in the base of $\mathcal{S}_h$), and $\mathbf{B}$ is the standard matrix relating displacements and strains. Then, the stiffness matrix can be computed as usual,

$$\mathbf{K} = \int_\Omega \mathbf{B}^T \mathbf{CB} \, d\Omega.$$

The modal analysis presented in the following is based on $\mathbf{K}$, which is naturally related to the energy norm in the finite dimensional interpolation space, $\mathcal{S}_h$, defined by the finite elements employed (and characterized by $\mathbf{B}$).

4.2. Comparing EFG and pseudo-divergence-free EFG

The incompressible limit is studied by evaluating the eigenvalues associated to each mode as the Poisson ratio, v, tends to 0.5. Following the procedure proposed by Huerta and Fernández-Méndez (2001) the logarithm of the eigenvalue is plotted as a function of the logarithm of $0.5 - v$. Then each mode is classified in three groups:

(1)   modes that do not present any locking behavior,
(2)   modes that do have physical locking, that is the eigenvalue goes to infinity because it is a volumetric mode, and
(3)   modes associated to non-physical locking, that is the eigenvalue goes to infinity but there is no volume variation.

In the last case, the displacement field conserves the total area but suffers from non-physical locking. The interpolation space is not rich enough to ensure the divergence-free condition.

In fact, these last modes do verify that

$$\int_\square \nabla \cdot u_h dx = 0,$$

but do not comply with the divergence-free condition locally (at each interior point). This is clearly a non-physical locking behavior.

The modal analysis is performed on a distribution of 3×3 particles and for bilinear consistency, that is $\mathbf{P} = \{1, x_1, x_2, x_1\, x_2\}^{\mathrm{T}}$. Figures 1 and 2 show the modes already classified for two different dilation parameters, $\rho/h = 1.2$ and 2.2.

Figure 3 compares the eigenvalues obtained by standard EFG and the pseudo-divergence-free interpolation for two particular non-physical locking modes. Moreover, three values of ratio ρ/h are also compared, namely 1.2, 2.2 and 3.2.

Note that the pseudo-divergence-free interpolation has not suppressed the non-physical locking modes. Thus, for a fixed dilation parameter ρ variations on the ratio ρ/h do not suppress locking. Indeed, the influence of locking is reduced because the eigenvalue is decreased. That is, the energy associated to the locking mode is decreased and this attenuates the volumetric locking. Nevertheless, in the incompressible limit, locking will still be present and it may induce useless numerical results.

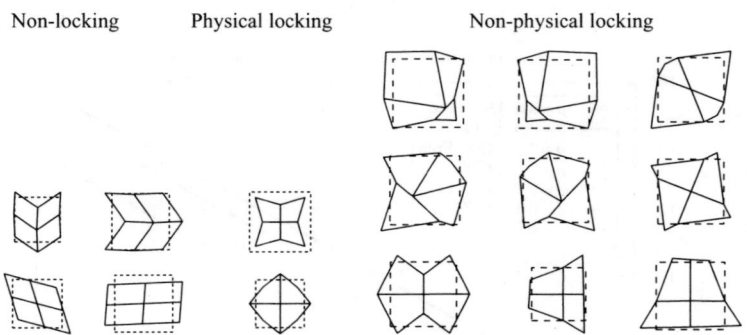

Figure 1 *Modes for a 3×3 distribution of particles with bilinear consistency and*
$\rho/h = 1.2$

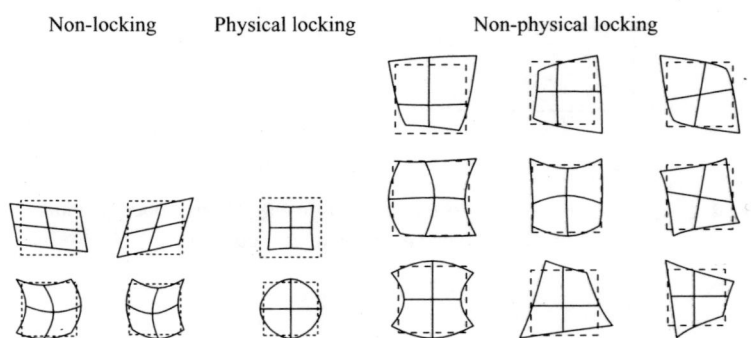

Figure 2 *Modes for a 3×3 distribution of particles with bilinear consistency and*
$\rho/h = 2.2$

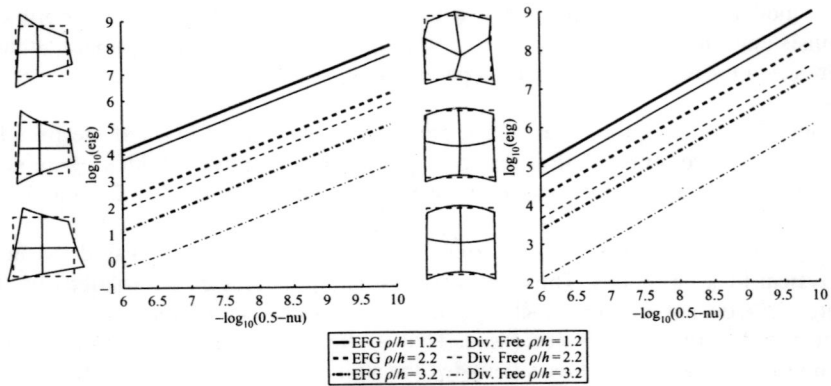

Figure 3 *Comparison between EFG and pseudo-divergence-free interpolations: variation of the eigenvalue as v goes to 0.5 for two non-physical locking modes with $\rho/h = 1.2, 2.2$ and 3.2*

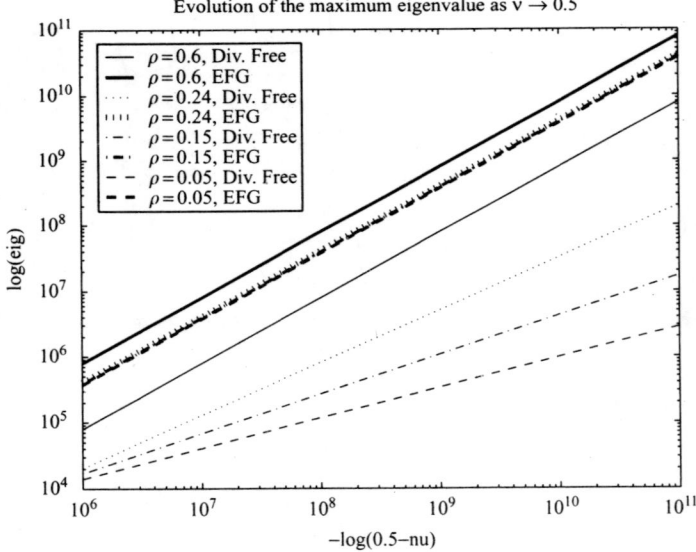

Figure 4 *Comparison between EFG and pseudo-divergence-free interpolations: variation of the maximum eigenvalue as v goes to 0.5*

These results should be expected. The convergence of the diffuse derivative, see (8), is ensured as ρ approaches zero for a ratio ρ/h kept constant. In other words, convergence is ensured as the interpolation is refined.

This is analyzed in Figure 4 for the non-physical locking mode that presents the largest eigenvalue (the first mode to spoil the approximation). These results are obtained for the "worst" dilation parameter, $\rho/h = 1.2$; that is, the one that induces results more similar to bilinear finite elements.

Four different values of ρ are tested, $\rho = 0.60, 0.24, 0.15$ and 0.05. It is important to note that as ρ decreases the eigenvalue also decreases (and drastically, the scale is logarithmic). Thus, as ρ decreases the influence of locking attenuates.

Moreover, and more importantly, the slope of the curve also decreases as ρ goes to zero (note that for standard EFG the slope remains constant). Thus, in the limit, as expected, the interpolation is divergence-free. Note that for $\rho = 0.05$ and $\nu = 0.5 - 10^{-11}$ the eigenvalue has been reduced in more than three orders of magnitude.

5. Numerical examples

5.1. The cantilever beam

As shown in Figure 5, a beam with linear isotropic material under plane strain conditions and with a parabolic traction applied to the free end is considered. This is a well-known example studied, for instance, by Hughes (2000) and Dolbow and Belytschko (1999). Displacements in both directions are prescribed at Γ_D. The prescribed displacements and the applied traction are such that the solution is known:

$$u_1 = -2\frac{1 - \nu^2}{E} y \left[(48 - 3x_1)x_1 + \left(2 + \frac{\nu}{1 - \nu}\right)\left(x_2^2 - 0.25\right) \right],$$

$$u_2 = 2\frac{1 - \nu^2}{E} \left[3\frac{\nu}{1 - \nu}x_2^2(8 - x_1) + \left(4 + 5\frac{\nu}{1 - \nu}\right)\frac{x_1}{4} + (24 - x_1)x_1^2 \right],$$

$$\sigma_{11} = -12x_2(8 - x_1), \quad \sigma_{22} = 0, \quad \sigma_{12} = 6\left(0.25 - x_2^2\right).$$

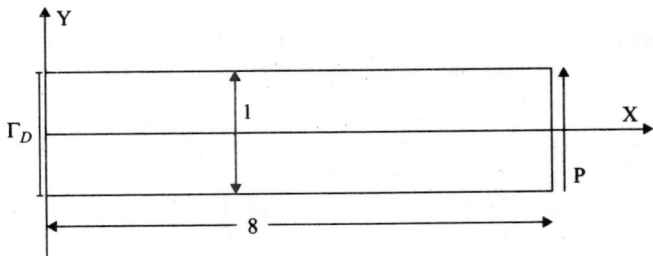

Figure 5 *Cantilever beam problem*

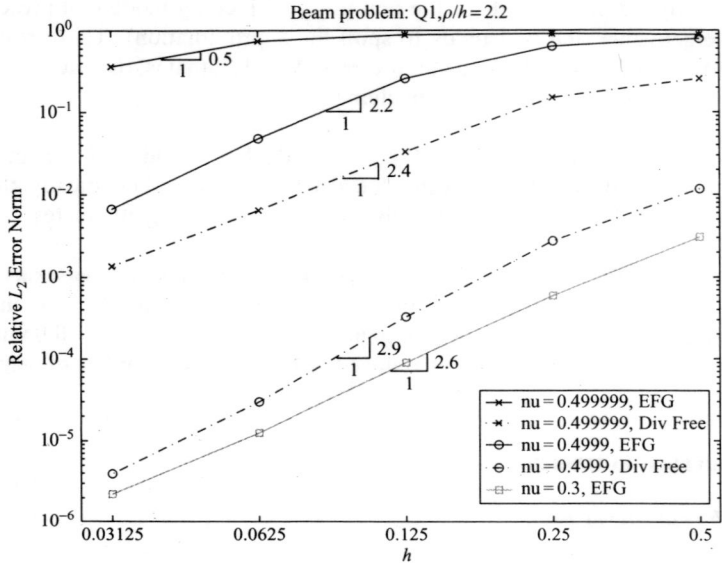

Figure 6 *Cantilever beam with bilinear consistency and $\rho/h = 2.2$*

The problem is solved with uniform distributions of particles. Figure 6 shows the relative L_2 error in displacements for $v = 0.3, 0.5 - 10^{-4}$ and $0.5 - 10^{-6}$. Results are shown for bilinear consistency and $\rho/h = 2.2$. The EFG results are compared with the pseudo-divergence-free interpolation. For EFG the typical convergence rates are obtained when $v = 0.3$, but, as expected, results degrade as v approaches the incompressible limit 0.5. However, the pseudo-divergence free interpolation is able to reproduce the theoretical rate of convergence even for a nearly incompressible case $v = 0.5 - 10^{-6}$ and a moderately fine discretization ($h < 0.25$, *i.e.* $\rho < 0.55$).

5.2. The plate with a hole

The stress field in an infinite plate with a hole subject to a far-field unit traction in the x direction is (Dolbow and Belytschko, 1999):

$$\sigma_{xx} = 1 - \frac{a^2}{r^2} \left(\frac{3}{2} \cos(2\theta) + \cos(4\theta) \right) + \frac{3a^4}{2r^4} \cos(4\theta)$$

$$\sigma_{yy} = -\frac{a^2}{r^2} \left(\frac{1}{2} \cos(2\theta) + \cos(4\theta) \right) - \frac{3a^4}{2r^4} \cos(4\theta)$$

$$\sigma_{xy} = -\frac{a^2}{r^2} \left(\frac{1}{2} \sin(2\theta) + \sin(4\theta) \right) + \frac{3a^4}{2r^4} \sin(4\theta)$$

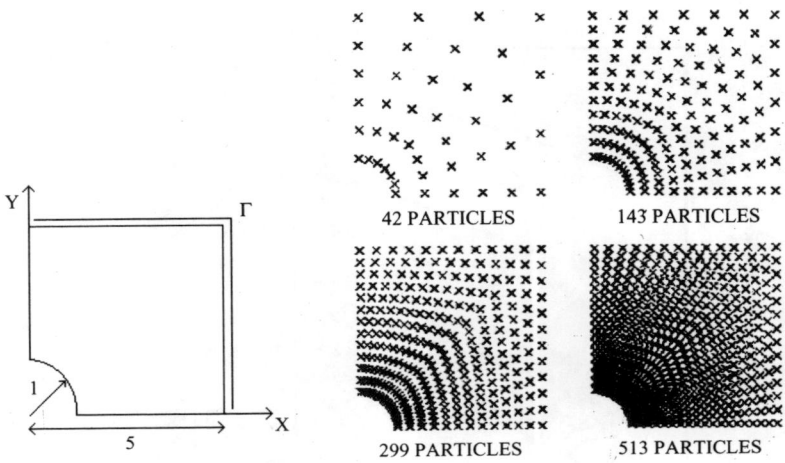

Figure 7 *Problem statement for the plate with hole and discretizations*

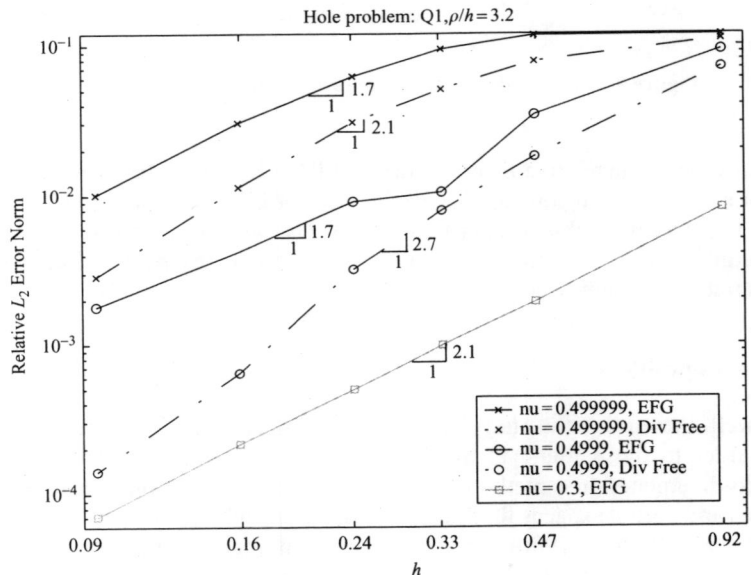

Figure 8 *Hole problem with bilinear consistency and $\rho/h = 3.2$*

where $a = 1$ is the hole radius, $r = \sqrt{x^2 + y^2}$ and $\theta = \arctan(y/x)$. The bounded upper quadrant shown in Figure 7 is used to solve the problem. Symmetry conditions are imposed in $x = 0$ and $y = 0$ and the tractions of the exact solution are considered in Γ.

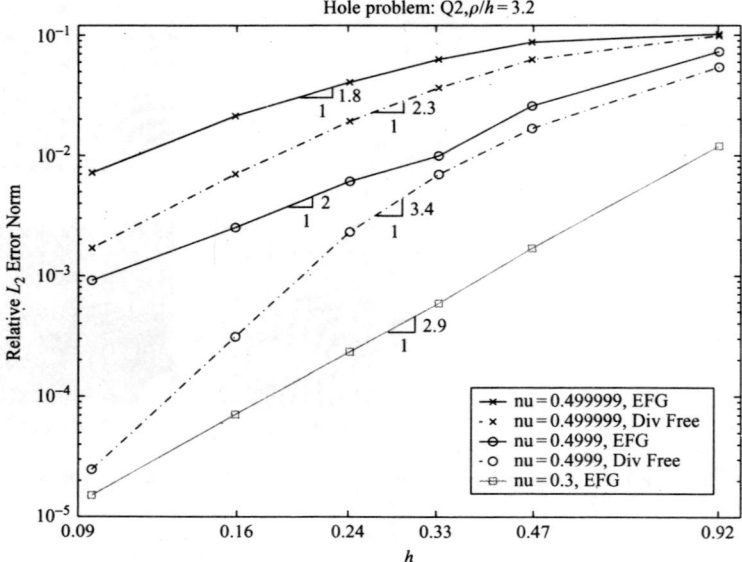

Figure 9 *Hole problem with biquadratic consistency and $\rho/h = 3.2$*

Figure 8 shows the relative L_2 error norm with $\nu = 0.3, 0.5 - 10^{-4}$ and $0.5 - 10^{-6}$. When $\nu = 0.3$ typical convergence is obtained for EFG but it suffers from locking when the incompressible limit is approached. The improved method maintains the convergence rate even with $\nu = 0.5 - 10^{-6}$. Similar results are obtained using biquadratic consistency. See Figure 9.

6. Stokes problem

It is well known that the study of viscous incompressible flows presents similar difficulties to those found in incompressible solid mechanics. Thus, here, the pseudo-divergence-free method is also used to solve the Stokes problem. Continuous and discrete spaces for Stokes equations are subject to an inf-sup condition (Girault and Raviart, 1986). This stability requirement is evidenced in practical computations by the existence of spurious pressure modes. The pseudo-divergence-free velocity field and the pressure field employed here comply with the LBB condition asymptotically.

6.1. *Statement of the problem*

Let Ω denote an open bounded region of $\mathbb{R}^2$ with boundary $\partial\Omega$. The 2D Stokes problem in Ω seeks a velocity field $\boldsymbol{u} = (u_1, u_2)$ and a pressure field p

such that:

$$\begin{cases} -\nu\Delta u + \nabla p = f & \text{in } \Omega, \\ \nabla \cdot u = 0 & \text{in } \Omega, \\ u = g & \text{on } \partial\Omega, \end{cases} \qquad (25)$$

where ν is the viscosity of the fluid and f is the body force (see Donea and Huerta, 2003).

6.2. Weak form

Given the problem defined in (25) with $u \in \mathcal{V}$ and $p \in \mathcal{Q}$, where $\mathcal{V} := [\mathcal{H}^1(\Omega)]^2$ and $\mathcal{Q} := \mathcal{L}_2(\Omega)$, the weak form of the Stokes problem, taking $g = 0$, is: find $u \in \mathcal{V}$, $p \in \mathcal{Q}$ such that

$$a(u, v) + b(v, p) + b(u, q) = (f, v) \quad \forall(v, q) \in \mathcal{V} \times \mathcal{Q},$$

where we define forms $a(\cdot, \cdot)$ and $b(\cdot, \cdot)$ as

$$a(u, v) := \int_\Omega \nabla v : \nu\nabla u \, d\Omega = \nu(\nabla u, \nabla v), \quad \text{and}$$

$$b(v, p) := -\int_\Omega p\nabla \cdot v \, d\Omega = -(p, \nabla \cdot v).$$

Note that $(\cdot, \cdot)$ denotes the standard $\mathcal{L}_2(\Omega)$-scalar product.

We now turn to consideration of an approximate discrete solution of the problem. Let $\mathcal{V}_\rho$ and $\mathcal{Q}_\rho$ denote finite dimensional subspaces of $\mathcal{V}$ and $\mathcal{Q}$ respectively. The index ρ refers to a characteristic measure of the support of the interpolation functions; it is related to the characteristic measure between particles, h. The discrete version of the problem, which in this case uses Nitsche's method (see Arnold, Brezzi, Cockburn and Marini, 2001/02; Babuska, Banerjee and Osborn, 2002; Becker, 2002; Stenberg, 1995) to impose boundary conditions, reads: find $u^\rho \in \mathcal{V}_\rho$, $p^\rho \in \mathcal{Q}_\rho$ such that, $\forall(v^\rho, q^\rho) \in \mathcal{V}_\rho \times \mathcal{Q}_\rho$,

$$a(u^\rho, v^\rho) + b(v^\rho, p^\rho) + b(u^\rho, q^\rho) - (\nu\partial_n u^\rho - p^\rho n, v^\rho)_{\partial\Omega}$$

$$- (u^\rho, \nu\partial_n v^\rho - q^\rho n)_{\partial\Omega} + \nu\frac{\gamma}{\rho}(u^\rho, v^\rho)_{\partial\Omega}$$

$$= (f, v^\rho) - (g, \nu\partial_n v^\rho - q^\rho n)_{\partial\Omega} + \nu\frac{\gamma}{\rho}(g, v^\rho)_{\partial\Omega}.$$

Now $(\cdot, \cdot)_{\partial\Omega}$ denotes the $\mathcal{L}_2(\partial\Omega)$-scalar product. Finally, the scalar γ is an arbitrary positive parameter that has to be chosen big enough in order to guarantee stability. Here an eigenvalue problem is solved as proposed by Griebel and Schweitzer (2002).

6.3. *Analytical test*

We consider a test problem with an analytical polynomial solution on the unit square, see Oden and Jacquotte (1984). Homogeneous Dirichlet boundary conditions are imposed on the whole boundary, and the theoretical rates of convergence, recall equation (8), shall be recovered numerically.

We consider the Stokes problem presented in (25) with $\Omega =]0, 1[\times]0, 1[$ and $g = 0$ on $\partial\Omega$; a polynomial force f is imposed in order to ensure the following solution of the problem:

$$\begin{cases} u_1(x, y) = x^2(1 - x)^2(2y - 6y^2 + 4y^3) \\ u_2(x, y) = (-2x + 6x^2 - 4x^3)y^2(1 - y)^2 \\ p(x, y) = x(1 - x) \end{cases}$$

We solve this problem with the pseudo-divergence-free MLS method and using $\rho/h = 1.2$ with a bilinear base to approximate both velocity and pressure.

The convergence results are shown in Figure 10. The velocity convergence rates for standard EFG and for the pseudo-divergence-free method are, as expected, similar. However, convergence in pressure is far from optimal in EFG, whereas it presents the theoretical slope in the proposed method. Recall that Eq (8) indicates that diffuse derivatives converge to the actual derivatives as $\rho \to 0$ ($\rho/h =$ cst). Since we use a bilinear base (*i.e.* $m = 1$) the convergence behaves as ρ^1. This means that if we double the number of particles (*i.e.*, if we divide ρ by 2) then the divergence must be at least divided by two. Figure 10 shows exactly this behavior.

6.4. *Driven cavity flow problem (leaky)*

Now we consider Stokes problem, equations (25), with $\Omega =]0, 1[\times]0, 1[$, $f = (0,0)^T$, $g = (0,0)^T$ on $\partial\Omega \setminus \{y = 1\}$ and $g = (1,0)^T$ on $\partial\Omega \cap \{y = 1\}$.

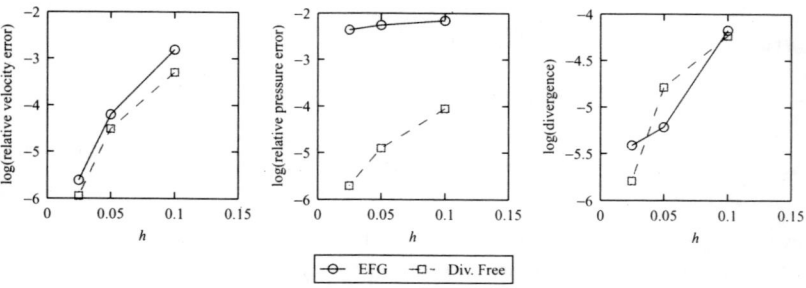

Figure 10 *Convergence results for velocities (left), pressures (middle) and divergence (right)*

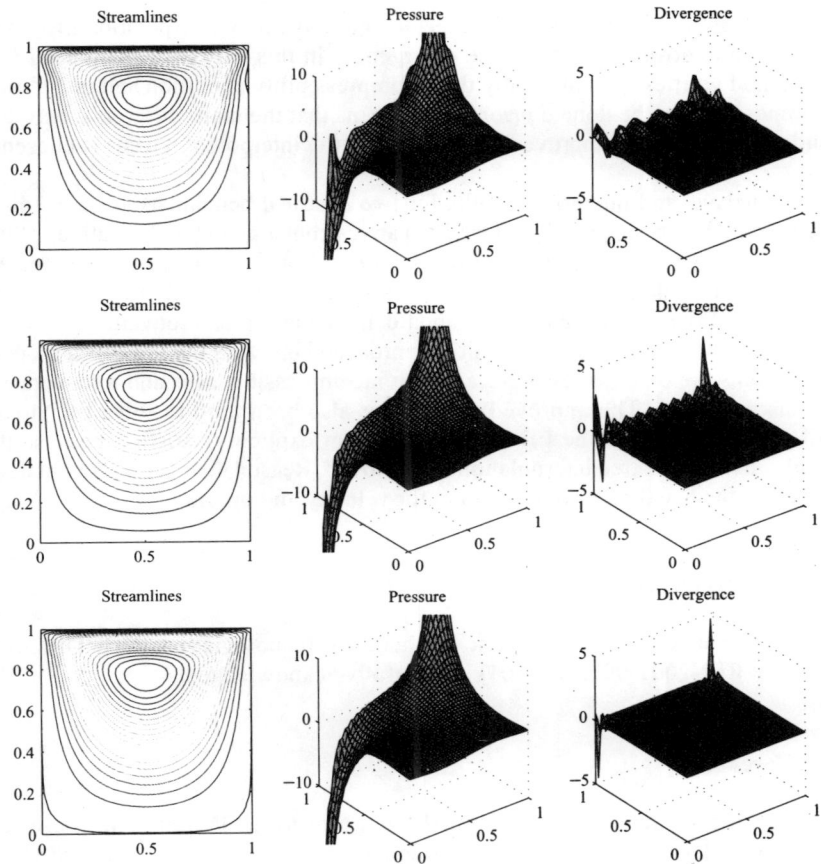

Figure 11 *Pseudo-divergence-free solution for a uniform distribution of* 11 × 11 *(top),* 21 × 21 *(middle) and* 41 × 41 *(bottom) particles.*

We solve this well-known benchmark problem with the pseudo-divergence-free method using $\rho/h = 2.1$ and a biquadratic base to approximate velocity and pressure. Streamlines, pressure distribution and divergence of u are depicted in Figure 11. Reasonable results are obtained in spite of the equal order interpolation for velocity and pressure. No spurious pressure modes are observed.

7. Conclusions

A novel improved formulation of the element-free Galerkin method is proposed in order to alleviate volumetric locking. It is based on a pseudo-divergence-free interpolation. Using the concept of diffuse derivatives as a convergence theorem of these derivatives to the ones of the exact solution, the new interpolation proposed is obtained imposing a zero diffuse divergence. In this way it is guaranteed that the method verifies asymptotically the incompressibility condition and in addition the imposition can be done *a priori*. This means that the main difference between standard EFG and the improved method is how the interpolation basis is chosen.

Modal analysis and numerical results for two classical benchmark tests in solids corroborate that, as expected, diffuse derivatives converge to the derivatives of the exact solution when the discretization is refined (for a fixed dilation parameter) and, of course, that diffuse divergence converges to the exact divergence with the expected theoretical rate. For standard EFG the typical convergence rate is degraded as the incompressible limit is approached but with the improved method good results are obtained even for a nearly incompressible case and a moderately fine discretization. The improved method has also been used to solve the Stokes equations. In this case the LBB condition is not explicitly satisfied because the pseudo-divergence-free interpolation is employed. Reasonable results are obtained in spite of the equal order interpolation for velocity and pressure.

Acknowledgements

The partial support of the Ministerio de Ciencia y Tecnología (contracts DPI2001-2204 and REN2001-0925-C03-01) is gratefully acknowledged.

8. References

Arnold, D. N., Brezzi, F., Cockburn, B. and Marini, L. D. (2001/02), 'Unified Analysis of Discontinuous Galerkin Methods for Elliptic Problems', *SIAM J. Numer. Anal.* **39**(5), 1749–1779.

Askes, H., de Borst, R. and Heeres, O. (1999), 'Conditions for locking-free Elastoplastic analyses in the Element-free Galerkin Method', *Comput. Methods Appl. Mech. Engng.* **173**, 99–109.

Babuska, I., Banerjee, U. and Osborn, J. E. (2002), *Meshfree Methods for Partial Differential Equations. Lecture Notes in Computational Science and Engineering*, Springer, chapter 'Meshless and Generalized Finite Element Methods: A Survey of Some Major Results'.

Babuska, I. and Suri, M. (1992), 'Locking Effects in the Finite Element Approximation of Elasticity problems', *Numerische Mathematik* **62**.

Becker, R. (2002), 'Mesh adaptation for Dirichlet Flow Control via Nitsche's Method', *Communications in Numerical Methods in Engineering* **18**, 669–680.

Belytschko, T., Krongauz, Y., Fleming, M., Organ, D. and Liu, W. K. (1996), 'Smoothing and Accelerated Computations in the Element-free Galerkin Method', *Journal of Computational and Applied Mathematics* **74**, 111–126.

Belytschko, T., Krongauz, Y., Organ, D., Fleming, M. and Krysl, P. (1996), 'Meshless Methods: an Overview and Recent Developments', *Comput. Methods Appl. Mech. Engng.* **139**, 3–47.

Belytschko, T., Liu, W. K. and Moran, B. (2000), *Nonlinear Finite Elements for Continua and Structures*, John Wiley & Sons Ltd., Chichester.

Belytschko, T., Lu, Y. Y. and Gu, L. (1994), 'Element-free Galerkin Methods', *Int. J. Numer. Meth. Engng.* **37**, 229–256.

Belytschko, T. and Tabarra, M. (1996), 'Dynamic Facture Using Element-free Galerkin methods', *Int. J. Numer. Meth. Engng.* **39**, 923–938.

Chen, J. S., Yoon, S.,Wang, H. and Liu,W. K. (2000), 'An Improved Reproducing Kernel Particle Method for Nearly Incompressible Finite Elasticity', *Comput. Methods Appl. Mech. Engng.* **181**, 117–145.

Dolbow, J. and Belytschko, T. (1999), 'Volumetric Locking in the Element-free Galerkin Method', *Int. J. Numer. Meth. Engng.* **46**, 925–942.

Donea, J. and Huerta, A. (2003), *Finite Element Methods for Flow Problems*, John Wiley & Sons Ltd., Chichester.

Duarte, C. A. M. and Oden, J. T. (1995), *Hp Clouds - a Meshless Method to Solve Boundary Value Problems*, Technical report, The University of Texas, Austin.

Duarte, C. A. M. and Oden, J. T. (1996), 'A h-p Adaptive Method Using Clouds', *Comp. Meth. Appl. Mech. Engng.* **139**(139), 237–262.

Girault, V. and Raviart, P.-A. (1986), *Finite Element Methods for Navier-Stokes Equations: Theory and Algorithms*, Springer-Verlag, Berlin.

Griebel, M. and Schweitzer, M. A. (2002), *A Particle-partition of Unity Method. Part v: Boundary Conditions*, Technical report, Institut fürAngewandte Mathematik, Universität Bonn.

Hermann, L. (1965), 'Elasticity Equations for Nearly Incompressible Materials by a Variational Theorem', *AIAA Journal* **3**, 1896–1900.

Huerta, A. and Fernández-Méndez, S. (2000), 'Enrichment and Coupling of the Finite Element and Meshless Methods', *Int. J. Numer. Meth. Engng.* **48**, 1615–1636.

Huerta, A. and Fernández-Méndez, S. (2001), 'Locking in the Incompressible Limit for the Element- free Galerkin Method', *Int. J. Numer. Meth. Engng.* **51**(11), 1361–1383.

Hughes, T. J. R. (2000), *The Finite Element Method: Linear Static and Dynamic Finite Element Analysis*, Dover Publications Inc., New York. Corrected reprint of the 1987 original [Prentice-Hall Inc., Englewood Cliffs, N.J.].

Liu, W. K., Belytschko, T. and Oden, editors, J. T. (1996), 'Meshless Methods', *Comput. Methods Appl. Mech. Engng.* **139**, 1–440.

Liu, W. K., Chen, Y., Jun, S., Chen, J. S., Belytschko, T., Pan, C., Uras, R. A. and Chang, C. T. (1996), 'Overview and Applications of the Reproducing Kernel Particle Methods', *Archives of Computational Methods in Engineering, State of the Art Reviews* **3**, 3–80.

Liu, W. K., Chen, Y., Uras, R. A. and Chang, C. T. (1996), 'Generalized Multiple Scale Reproducing Kernel Particle Methods', *Comput. Methods Appl. Mech. Engng.* **139**, 91–158.

Liu, W. K., Jun, S., Adee, J. and Belytschko, T. (1995), 'Reproducing Kernel Particle Methods for Structural Dynamics', *Int. J. Numer. Meth. Engng.* **38**, 1655–1679.

Liu, W. K., Jun, S. and Zhang, Y. F. (1995), 'Reproducing Kernel Particle Methods', *Int. J. Numer. Meth. Fluids* **20**, 1081–1106.

Liu, W. K., Li, S. and Belytschko, T. (2000), 'Moving least square reproducing kernel methods. (i) methodology and convergence', *Comput. Methods Appl. Mech. Engng.* **143**, 113–154.

Lu, Y. Y., Belytschko, T. and Gu, L. (1994), 'A New Implementation of the Elment-free Galerkin Method', *Comp. Meth. Appl. Mech. Engng.* **113**(113), 397–414.

Lucy, L. (1977), 'A Numerical Approach to Testing the Fission Hypothesis', *Aston. J.* **82**, 1013–1024.

Melenk, J. M. and Babuska, I. (1996), 'The Partition of Unity Finite Element Method: Basic theory and Applications', *Comput. Methods Appl. Mech. Engng.* **139**, 289–314.

Monaghan, J. J. (1988), 'An Introduction to sph', *Comput.Phys. Commun.* **48**, 89–96.

Nayroles, B., Touzot, G. and Villon, P. (1992), 'Generating the fFnite Element Method: Diffuse and Diffuse Elements', *Computational Mechanics* **10**, 307–318.

Oden, J. T. and Jacquotte, O. P. (1984), 'Stability of Some Mixed Finite Element Methods for Stokesian Flows', *Comput. Methods Appl. Mech. Engng.* **43**, 231–247.

Randles, P. W. and Libersky, L. D. (1996), 'Smoothed Particle Hydrodynamics: Some Recent Improvements and Applications', *Comp. Meth. Appl. Mech. Engng.* **139**, 375–408.

Stenberg, R. (1995), 'On Some Techniques for Approximating Boundary Conditions in the Finite Element Method', *J. Comp. Appl. Maths.* **63**, 139–148.

Suri, M. (1996), 'Analytic and Computational Assessment of Locking in the hp Finite Element Method', *Comput. Methods Appl. Mech. Engng.* **133**, 347–371.

Villon, P. (1991), *Contribution à l'optimisation*, Thèse présentée pour l'obtention du grade de docteur d'état, Université de Technologie de Compiègne, Compiègne, France.

Zhu, T. and Atluri, S. N. (1998), 'A Modified Collocation Method and a Penalty Formulation for Enforcing the Essential Boundary Conditions in the Element-free Galerkin Method', *Computational GMechanics* **21**, 211–222.

Chapter 3

Alternative Total Lagrangian Formulations for Corrected Smooth Particle Hydrodynamics (CSPH) Methods in Large Strain Dynamic Problems

Javier Bonet & Sivakumar Kulasegaram
Department of Civil Engineering, University of Wales Swansea, Swansea SA2 8PP, UK

1. Introduction

Smooth particle hydrodynamics (SPH) is a truly meshfree, simple and robust computational technique that can be used in numerical simulations of various engineering problems. The method was pioneered in 1977 for modelling astrophysical and cosmological problems, and since the early 90's the application has been extended to numerous areas of computational mechanics [1–10]. However, in spite of its attraction for computational mechanics, the method suffers from lack of accuracy and more importantly instability from lack of nodal completeness and/or integrability of the approximations for functions and their derivatives. Recently, a number of techniques have been developed to circumvent such difficulties encountered in SPH and other meshless methods [10–15].

One of the major obstacles generally faced in meshless methods such as SPH is the presence of tensile instability in the formulation of solid applications [11–15]. It has been reported that tensile instability is to a large extent associated with using Eulerian kernels [12], where the derivatives of the kernel functions are constantly changing as the particles move. The changes in the internal forces brought about by these changes in the derivatives of the Eulerian kernels give rise to spurious terms in the tangent stiffness of the system which are the main cause of so called tension instability [12]. A number of techniques have been developed to address this issue in the case of SPH and related meshfree methods [11,12,15]. One such technique is based on formulating Lagrangian continuum equations whereby the internal forces are evaluated with respect to a fixed reference configuration. In this case the kernel function and its derivatives are based at the reference configuration and hence do not depend on the current position of the particles [11,12]. Thus the tensile instability will be completely eliminated or transformed into spurious mechanisms, which can be easily controlled by the use of artificial viscosity. The detailed analysis of tensile instability and alternative approaches to eliminate these instabilities can be found in the literature [11–15]. This paper mainly dwells on two different ways of formulating SPH in a Lagrangian setting and compares their salient features. Several numerical examples are presented to demonstrate the ability of the formulations to simulate complex problems.

2. Numerical methodology

2.1. *SPH approximations*

In meshfree methods such as SPH, any problem variable and its gradient are generally interpolated from values at a discrete number of particles by using the following approximations:

$$\phi(x) = \sum_{b=1}^{N} V_b \phi_b W_b(x) \tag{1}$$

$$\nabla\phi(x) = \sum_{b=1}^{N} V_b \phi_b \nabla W_b(x) \tag{2}$$

where V_b denotes the volume of material associated with a given particle and W_b represents the 'kernel' or interpolation function, which usually has a bell shape with a compact support as shown in Figure 1.

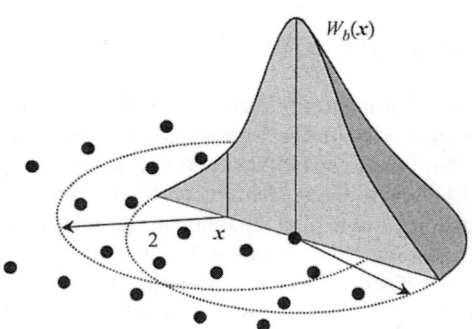

Figure 1 *Particle interpolation and kernel function*

The most commonly used kernel function in SPH is a cubic spline kernel function given by,

$$W(x) = \frac{c}{h^d} \begin{cases} 1 - \frac{3}{2}\xi^2 + \frac{3}{4}\xi^3 & \text{if } \xi \leq 1 \\ \frac{1}{4}(2-\xi)^3 & \text{if } 1 < \xi \leq 2; \xi = \|x\|/h \\ 0 & \text{if } \xi > 2 \end{cases} \tag{3}$$

where d is the number of dimensions of the problem and c is a scaling factor to normalise the kernel function. Here, the length parameter h has a similar interpretation to the element size in the finite element method. For instance, applying equation (1) to density of a continuum leads to the classical SPH equation:

$$\rho(x) = \sum_{b=1}^{N} m_b W_b(x) \tag{4}$$

In this way, the SPH representation of the governing equations can be built from fundamental equations of motion.

2.2. Continuum equations

Consider a 3-dimensional continuum shown in Figure 2 undergoing a given motion defined by a mapping ϕ between initial and current position as:

$$x = \phi(X, t) \tag{5}$$

which gives the position x of each material point X as a function of time. The deformation gradient F is defined as:

$$F = \nabla_0 \phi = \frac{\partial x}{\partial X} \tag{6}$$

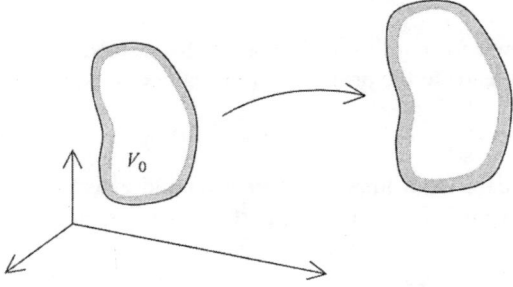

Figure 2 *Continuum deformation*

and the Jacobian or volume ratio J of the continuum is given by:

$$J = \det F = \frac{dV}{dV_0} \tag{7}$$

where dV_0 and dV represent the initial and the current element volumes. Conservation of mass or continuity equation is then obviously expressed as;

$$\rho_0 = \rho J \tag{8}$$

where ρ_0 and ρ are initial and current densities of volume elements.

The momentum balance equation for the deformable body is given by,

$$\rho_0 a = \rho_0 f_0 + \nabla_0 P : I \tag{9}$$

where a is acceleration, f_0 is body force per unit mass, typically due to gravity g and P is first Piola-Kirchhoff tensor. The first Piola-Kirchoff tensor can be expressed in terms of Cauchy stress tensor as,

$$P = J\sigma F^{-T} \tag{10}$$

And finally, in the absence of heat transfer, the conservation of energy of the continuum can be written by,

$$\rho_0 \dot{E} = P : \dot{F} \tag{11}$$

where E is the internal energy per unit mass. As the analyses in this paper will be confined to isothermal processes the terms corresponding to heat energy have been ignored in the above energy conservation equation. The following section deals with discretization of the governing equations based on Lagrangian SPH formulations.

2.3. Discrete equilibrium equations

In order to discretise the equilibrium equation (9), a standard Galerkin approach is employed, which leads to the principle of virtual work expressed as,

$$\delta \dot{w}_{\text{ine}} = \delta \dot{w}_{\text{ext}} - \delta \dot{w}_{\text{int}} \tag{12}$$

where using standard mass lumping the inertia and external (gravity due) virtual work rates are expressed in terms of particle masses, accelerations and virtual velocities as:

$$\delta \dot{w}_{\text{ine}} = \sum_I m_I a_I \cdot \delta v_I; \quad \delta \dot{w}_{\text{ext}} = \sum_I m_I g \cdot \delta v_I \tag{13}$$

and the internal virtual work will be expressed in following sections in terms of the internal equivalent forces as,

$$\delta \dot{w}_{\text{int}} = \int_{V_0} \boldsymbol{P} : \delta \dot{\boldsymbol{F}} \, dV_0 = \sum_I \boldsymbol{T}_I \cdot \delta \boldsymbol{v}_I \tag{14}$$

The standard dynamic equilibrium equation for a given particle is then obtained as

$$m_I a_I = m_I g - T_I \tag{15}$$

3. Lagrangian SPH

Previous research has revealed that the discretisation of the continuum equations based on the framework of a fully Lagrangian formulation eliminates undesirable effects due to tensile instabilities [12]. In Lagrangian SPH all derivatives are taken with respect to a constant reference configuration where both the kernel and its derivatives remain constant. In this section two different Lagrangian formulations, one based on corrected SPH formulation and another based on traditional SPH formulation are described.

3.1. Corrected SPH formulation

Consider a general deformation of a body discretised using a number of SPH particles as shown in Figure 3. The deformation gradient defined by equation (6) can now be evaluated at a given particle I in terms of the current particle positions as:

$$\boldsymbol{F}_I = \nabla_0 \phi = \sum_{J=1}^{N} \boldsymbol{x}_J \otimes \boldsymbol{G}_J(\boldsymbol{X}_I) \tag{16}$$

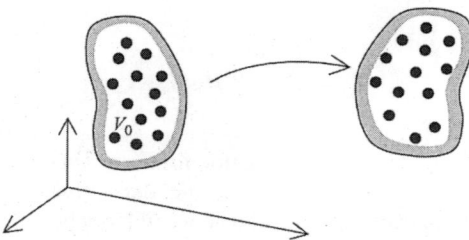

Figure 3 *General deformation from reference to current*

where the gradient functions G contain the corrected kernel gradients $\tilde{\nabla} W(X)$ at the initial reference configuration, that is:

$$G_K(X_I) = V_K \tilde{\nabla}_0 W_K(X_I) \tag{17}$$

Possible ways of correcting the above kernel or its gradient that will ensure linear completeness are discussed in references [9–11]. Note that the necessary correction will be performed at the initial reference configuration of the body.

In order to derive general equations for the internal forces using a Lagrangian corrected SPH technique, consider the internal virtual work equation expressed in the reference configuration in terms of the first Piola-Kirchoff tensor P as,

$$\delta \dot{w}_{int} = \int_{V_0} P : \delta \dot{F} \, dV_0 = \sum_{J=1}^{N} V_J^0 P_J : \delta \dot{F}_J \tag{18}$$

The virtual deformation gradient rate emerges from equation (16) as:

$$\delta \dot{F}_J = \sum_{K=1}^{N} \delta v_K \otimes G_K(X_J) \tag{19}$$

which, upon substitution into equation (18) yields the virtual internal work expression as:

$$
\begin{aligned}
\delta \dot{w}_{int} &= \sum_{J=1}^{N} V_J^0 P_J : \delta \dot{F}_J \\
&= \sum_{J=1}^{N} V_J^0 P_J : \left(\sum_{K=1}^{N} \delta v_K \otimes G_K(X_J) \right) \\
&= \sum_{K=1}^{N} \delta v_K \cdot \left(\sum_{J=1}^{N} V_J^0 P_J G_K(X_J) \right)
\end{aligned} \tag{20}
$$

thus enabling the internal force vector corresponding to a given node I to be easily identified from this expression as:

$$T_I = \sum_{J=1}^{N} V_J^0 P_J G_I(X_J) \tag{21}$$

The obvious advantage of using this equation for the evaluation of internal forces in a discretized continuum is the fact that the kernel derivative functions G are fixed at the initial (or reference) configuration and hence will not be dependent upon current nodal positions. In addition it can be noted that the corrections are also evaluated at the initial configuration thus enabling one to reduce the computational cost.

3.2. *Alternative SPH formulation*

The previous formulation relies on linearly corrected kernel gradient vectors obtained in the initial configuration. An alternative simpler formulation, which leads to equations similar to the traditional SPH, is presented in this section.

Consider first the approximation of the deformation gradient tensor $\boldsymbol{F}$. Given that this is a two point tensor relating initial vectors $d\boldsymbol{X}$ to their final counterparts $d\boldsymbol{x} = \boldsymbol{F} d\boldsymbol{X}$, a simple approximation at a given particle I is given by the weighted average of dyadic (tensor) products of incremental vectors as:

$$\boldsymbol{F}_I = \left(\sum_{J=1}^{N} m_I W_{IJ} (\boldsymbol{x}_J - \boldsymbol{x}_I) \otimes (\boldsymbol{X}_J - \boldsymbol{X}_I) \right) \boldsymbol{M}_I^{-1} \tag{22}$$

where W_{IJ} denotes the uncorrected derivative of the kernel with respect to particle distance as,

$$W_{IJ} = \frac{1}{r_{IJ}} \frac{dW_I(\boldsymbol{X}_J)}{dr_{IJ}}; \quad r_{IJ}^2 = (\boldsymbol{X}_J - \boldsymbol{X}_I) \cdot (\boldsymbol{X}_J - \boldsymbol{X}_I) \tag{23}$$

and the correction matrix $\boldsymbol{M}$ is chosen so that for the case with uniform deformation gradient, where $\boldsymbol{x}_J - \boldsymbol{x}_I = \boldsymbol{F}(\boldsymbol{X}_J - \boldsymbol{X}_I)$, the exact deformation gradient is found. A simple substitution shows that this matrix must consequently be given by:

$$\boldsymbol{M}_I = \sum_{J=1}^{N} W_{IJ} m_J (\boldsymbol{X}_J - \boldsymbol{X}_I) \otimes (\boldsymbol{X}_J - \boldsymbol{X}_I) \tag{24}$$

In order to derive the corresponding equations for internal forces based on the above alternative formulation, consider again the internal virtual work equation as:

$$\delta \dot{w}_{\text{int}} = \sum_{I=1}^{N} m_I \frac{\boldsymbol{P}_I}{\rho_I^0} : \delta \dot{\boldsymbol{F}}_I \tag{25}$$

where the virtual deformation gradient rate is now derived from equation (22) to give:

$$\delta \dot{\boldsymbol{F}}_I = \left(\sum_{J=1}^{N} W_{IJ} m_J (\delta \boldsymbol{v}_J - \delta \boldsymbol{v}_I) \otimes (\boldsymbol{X}_J - \boldsymbol{X}_I) \right) \boldsymbol{M}_I^{-1} \tag{26}$$

Substituting the above equation into the virtual work expression gives

$$\delta \dot{w}_{\text{int}} = \sum_{I,J=1}^{N} W_{IJ} m_I m_J \frac{\boldsymbol{P}_I \boldsymbol{M}_I^{-1}}{\rho_I^0} : (\delta \boldsymbol{v}_J - \delta \boldsymbol{v}_I) \otimes (\boldsymbol{X}_J - \boldsymbol{X}_I) \tag{27}$$

Using simple algebra the above expression can be re-written as:

$$\delta \dot{w}_{\text{int}} = \sum_{K=1}^{N} \delta \boldsymbol{v}_K \cdot \left(\sum_{L=1}^{N} m_K m_L W_{LK} \frac{\boldsymbol{P}_L \boldsymbol{M}_L^{-1}}{\rho_L^0} + m_K m_L W_{KL} \frac{\boldsymbol{P}_K \boldsymbol{M}_K^{-1}}{\rho_K^0} \right) (\boldsymbol{X}_K - \boldsymbol{X}_L)$$

(28)

thus enabling the internal force vector corresponding to a given node I to be easily identified from this expression as:

$$\boldsymbol{T}_I = \sum_{J=1}^{N} m_I m_J \left(W_{JI} \frac{\boldsymbol{P}_J \boldsymbol{M}_J^{-1}}{\rho_J^0} + W_{IJ} \frac{\boldsymbol{P}_I \boldsymbol{M}_I^{-1}}{\rho_I^0} \right) (\boldsymbol{X}_I - \boldsymbol{X}_J)$$

(29)

For the particular case where kernel functions are symmetric that is,

$$W_{IJ} = -W_{JI} \quad \text{and} \quad \nabla_0 W_I(\boldsymbol{X}_J) = -\nabla_0 W_J(\boldsymbol{X}_I) = (\boldsymbol{X}_I - \boldsymbol{X}_J) W_{IJ}$$

(30)

the internal force $\boldsymbol{T}_I$ can be written as,

$$\boldsymbol{T}_I = \sum_{J=1}^{N} m_I m_J \left(\frac{\boldsymbol{P}_J \boldsymbol{M}_J^{-1}}{\rho_J^0} + \frac{\boldsymbol{P}_I \boldsymbol{M}_I^{-1}}{\rho_I^0} \right) \nabla_0 W_I(\boldsymbol{X}_J)$$

(31)

As the matrices $\boldsymbol{M}$ are fixed at the initial (or reference) configuration, the above internal force equation shares the same advantages of equation (21). In addition, the final equation for the internal forces closely resembles the equations used in standard SPH.

3.3. Preservation of momentum

It is essential to examine whether the Lagrangian formulations proposed in the above discussions satisfy the momentum preservation conditions usually required of a continuum equation. For this purpose internal force equations linear and angular momentum of the Lagrangian formulations are analysed. It can be recalled that linear momentum is preserved whenever the sum of the internal forces of each particle vanishes for any state of stresses, that is:

$$\sum_{I=1}^{N} \boldsymbol{T}_I = \boldsymbol{0}$$

(32)

In the case of corrected SPH formulation substituting for $\boldsymbol{T}_I$ from equation (21) into the above condition gives, after simple algebra,

$$\sum_{I=1}^{N} \boldsymbol{T}_I = \sum_{J=1}^{N} V_J \boldsymbol{P}_J \left(\sum_{I=1}^{N} \boldsymbol{G}_I(\boldsymbol{X}_J) \right)$$

(33)

From the above equation it can be stated that the conservation of linear momentum enforces the following requirement on the initial gradient vectors G:

$$\sum_{I=1}^{N} G_I(X_b) = 0 \tag{34}$$

This is simply the order zero completeness condition which ensures that the gradient of a constant function vanishes. A number of different techniques to ensure that this condition is satisfied are reported in the literature [10–12].

It is trivial to prove that the alternative SPH formulation defined in section 3.2 satisfies equation (32) and hence preserves linear momentum. In general, internal force at particle I can be expressed as the sum of interaction forces between pairs of particles as (see Figure 4)

$$T_I = \sum_{J=1}^{N} T_{IJ} \tag{35}$$

For instance, if the internal forces are given by equation (31), then the interaction force is

$$T_{IJ} = m_I m_J \left(\frac{P_I M_I^{-1}}{\rho_I^0} + \frac{P_J M_J^{-1}}{\rho_J^0} \right) \nabla_0 W_J(X_I) \tag{36}$$

Given that $\nabla W_I(X_J) = -\nabla W_J(X_I)$, it is clear that $T_{IJ} = -T_{JI}$, and consequently the total sum of all interaction pairs will vanish (see Figure 4).

Although most of the formulations will preserve linear momentum, the same is not true for angular momentum. It can be recalled that the angular momentum is preserved when the total moment of the internal forces with respect to an arbitrary reference point vanishes, that is:

$$\sum_{I=1}^{N} x_I \times T_I = 0 \tag{37}$$

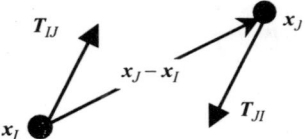

Figure 4 *Interaction forces between two particles*

In the case of corrected SPH formulation, again introducing the internal force equation (21) gives after simple algebra:

$$\sum_{I=1}^{N} x_I \times T_I = -\varepsilon : \sum_{I=1}^{N} T_I \otimes x_I$$

$$= -\varepsilon : \sum_{J=1}^{N} V_J^0 P_J \left(\sum_{I=1}^{N} G_I(X_J) \otimes x_I \right) \qquad (38)$$

where ε denotes the third order alternating tensor. The term in brackets in the above equation coincides with the transpose of the deformation gradient tensor given by equation (16). Taking this into account together with the equation (10) relating Cauchy and Piola-Kirchhof stresses and the symmetry of σ gives:

$$\sum_{I=1}^{N} x_I \times T_I = -\varepsilon : \sum_{J=1}^{N} V_J^0 P_J F_J^T$$

$$= -\varepsilon : \sum_{J=1}^{N} V_J \sigma_J$$

$$= 0 \qquad (39)$$

Consequently, internal force equation (21) preserves angular momentum for any choice for initial gradient vectors G.

In the case of the alternative Lagrangian SPH formulation discussed in section 3.2, a similar derivation, again relying on the symmetry of the Cauchy stresses and equation (22), enables the preservation of angular momentum to be proved as,

$$\sum_{I=1}^{N} x_I \times T_I = -\varepsilon : \sum_{I=1}^{N} T_I \otimes x_I$$

$$= -\varepsilon : \sum_{I,J=1}^{N} m_I m_J \left(W_{JI} \frac{P_J M_J^{-1}}{\rho_J^0} + W_{IJ} \frac{P_I M_I^{-1}}{\rho_I^0} \right) (X_I - X_J) \otimes x_I$$

$$= -\varepsilon : \sum_{I,J=1}^{N} m_I m_J \left(W_{JI} \frac{P_J M_J^{-1}}{\rho_J^0} \right) (X_I - X_J) \otimes (x_I - x_J)$$

$$= -\varepsilon : \sum_{J=1}^{N} m_J P_J F_J^T$$

$$= -\varepsilon : \sum_{J=1}^{N} m_J \sigma_J$$

$$= 0 \qquad (40)$$

4. Numerical examples

In order to illustrate the ability of Lagrangian SPH formulations various numerical examples are presented in this section. The examples are targeted to demonstrate the simulations of large strain three dimensional problems involving elasto-plastic and hyperelastic materials.

4.1. *Taylor Bar Impact*

This section presents numerical results from the simulation of a small cylindrical copper bar against a rigid planar wall. The bar has an initial length of 0.0324 m and initial radius 0.0032 m. The initial velocity of the bar is 227 m/s and the termination time of the problem is 80 μs. Von Mises plasticity with linear isotropic hardening is employed for the numerical computation. Material properties used for copper are given in Table 1 [17]. This is a classical dynamic test example and the results obtained match closely with those achieved using a standard FE formulation [17].

Figure 5 shows the deformed shape and the distribution of equivalent plastic strain at various stages of the numerical simulation. The results show that the meshless method yields larger maximal equivalent plastic strain than finite elements, but the discrepancy is small.

4.2. *Bending of hyperelastic cylinder*

This example involves a nearly incompressible neo-Hoookean cylinder travelling with initial velocity of 1.88 m/s to the right which is suddenly fixed at its base. The initial radius is 0.32 m and the length 3.24 m. The shear modulus is 0.3571 MN/m^2 and the bulk modulus is 1.67 MN/m^2. The shapes obtained at different times are shown in Figure 6. The same example has been run using a standard dynamic FE code with identical initial nodal positions and tri-linear 8 node cube elements. The SPH and FE solution for the centreline of the cylinder at three different times are compared in Figure 7, where the agreement can be seen to be excellent.

Table 1 *Copper bar material properties*

Elastic Modulus E	117 GN/m^2
Poisson's Ratio ν	0.35
Yield Stress σ_y	0.4 GN/m^2
Hardening Modulus H	0.1 GN/m^2
Density ρ	8930 kg/m^3

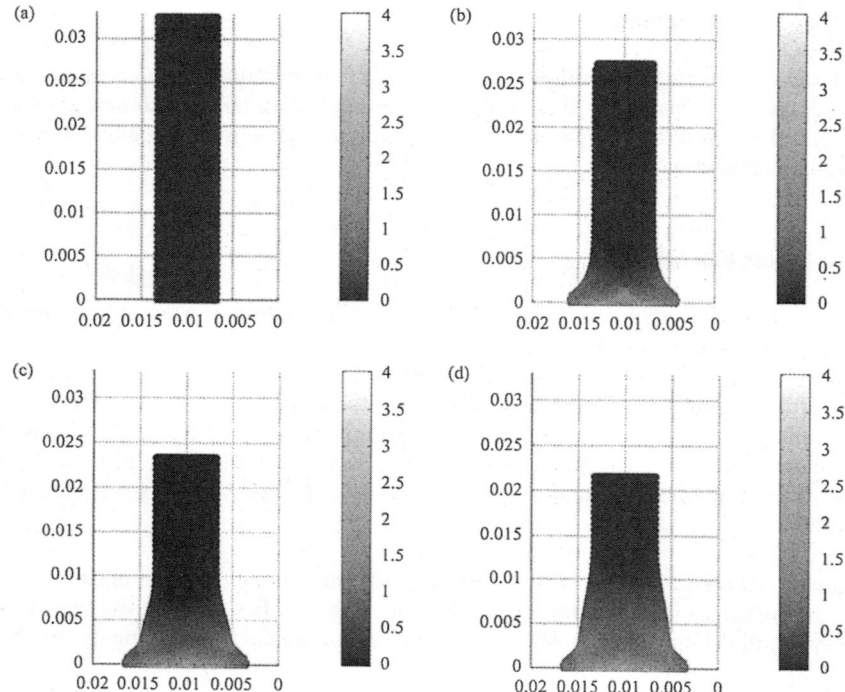

Figure 5 *Deformed shape and equivalent plastic strain of a Taylor bar at various stages*

4.3. *High speed impact of brittle materials*

In this section the high speed impact and fracture of tungsten cube is simulated using Lagrangian SPH formulations. A silica-phenolic target panel is impacted at right angles with the tungsten cube as described in Figure 8. The numerical computation is performed for a 42.2 g tungsten cube travelling at 1930 m/s before striking a long stationary bar made of silica-phenolic material.

In addition to the governing equations discussed in section 2.2, the pressure equation incorporating specific internal energy is used in the computation. For this purpose, the pressure is evaluated by using Mie-Gruneisen equation of state given as [6,18],

$$P = \left(1 - \frac{1}{2}\Gamma\mu\right) P_H(\rho) + \Gamma\rho E; \quad \mu = \frac{\rho}{\rho_0} - 1 \tag{41}$$

where Γ is Gruneisen material parameter and P_H is Hugnoit pressure given by,

$$P_H(\rho) = \frac{C^2\mu(1+\mu)}{[1 - \mu(S-1)]^2} \tag{42}$$

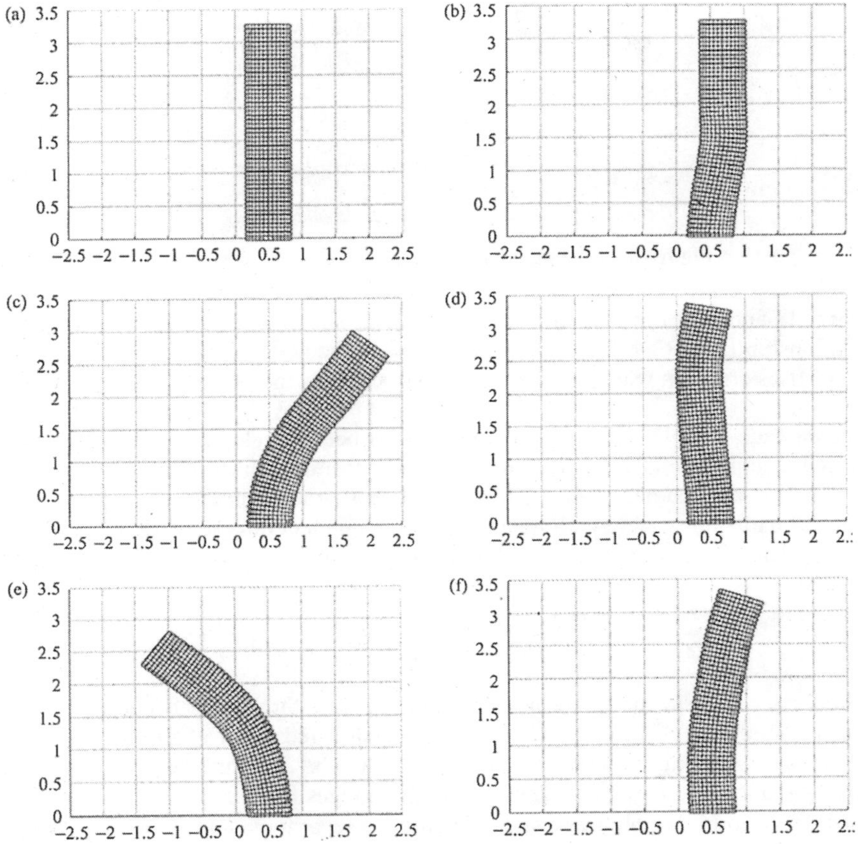

(a)

(b)

(c)

(d)

(e)

(f)

Figure 6 *Deformed shapes of the hyperelastic material at various stages*

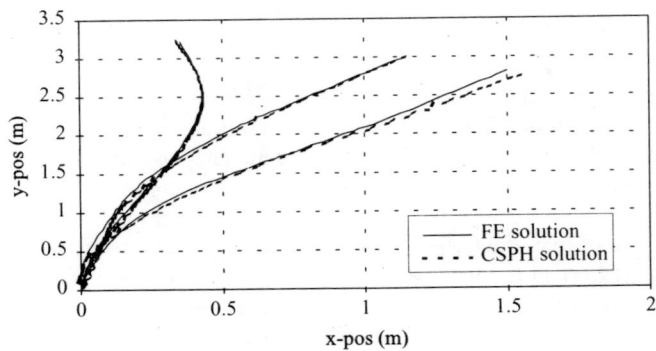

Figure 7 *FE vs. CSPH comparison for the oscillating cylinder*

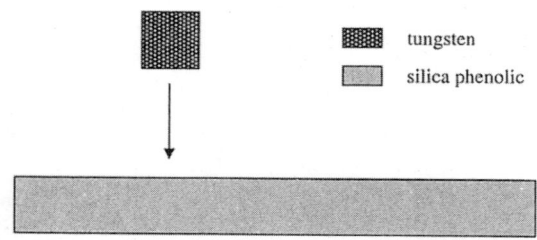

tungsten

silica phenolic

Figure 8 *Schematic diagram of the numerical experiment*

where C and S are the parameters in the linear shock velocity–particle velocity relationship $U_s = C + SU_p$. The plastic flow of the material is determined by the Von Mises criterion when the stress invariant exceeds the yield strength.

The damage model used in this simulation has been developed by Randles *et al.* [18,19] and is generally considered suitable for treating impact fracture of tungsten which may be a combination of ductile and brittle mechanisms. This model uses a scalar damage variable without attempting to connect directly with microscopic mechanisms. The evolution of the scalar damage variable D with time is postulated as [18],

$$\frac{dD}{dt} = \frac{1}{\tau}\left[\frac{(\sigma_{\max} - \sigma_{th})}{\sigma_{th0}}\right]^2 \left(\frac{1}{1 - D^2}\right) \qquad \sigma_{\max} > \sigma_{th} \qquad (43)$$

where $\sigma_{\max}$ is the maximum principal stress, σ_{th} is the threshold stress for the onset of tensile damage, σ_{th0} is the threshold for undamaged material and τ is the time constant controlling the rate of damage growth. The accumulation of damage is zero when $\sigma_{\max} < \sigma_{th0}$. The damage variable ranges from 0 to 1 with 0 denoting no damage and 1 complete damage with possible separation, and values in between denoting various states of damage between undamaged material and complete separation. The effects of damage evolution on material properties are given by [18],

$$\sigma_{th} = \sigma_{th0}(1 - D^2) \quad k = k_0(1 - D^2) \quad Y = Y_0(1 - D^2) \quad G = G_0(1 - D^2)$$
$$(44)$$

where k is tensile bulk modulus, Y is the yield stress for plastic deformation, and G is shear modulus. The k_0, Y_0 and G_0 are initial values of the corresponding material properties. The detail analysis of this damage model and its implementation can be found in the reference [18,19].

The properties of materials used in the numerical simulations are listed in Table 2 [18]. Figures (9.Ia) and (9.Ib) show the initial configurations of side and top views of the materials. And the Figures (9.IIa) and (9.IIb) show the views of the materials after the complete penetration of the tungsten cube. The above initial set up is chosen in order to reduce the computational effort required. In the present example a penalty based contact-impact algorithm has been adopted [20,21].

Table 2 *Material properties of tungsten and silica phenolic*

Material	ρ_0(g/cc)	C (mm/μs)	S	G_0(Mb)	Y_0 (Kb)	Γ	σ_{th0} (Kb)	$\tau(\mu$s)
Tungsten	19.23	4.00	1.23	1.540	60.0	1.54	35.00	0.05
Silica phenolic	1.71	3.24	1.39	0.038	10.0	1.00	2.00	0.50

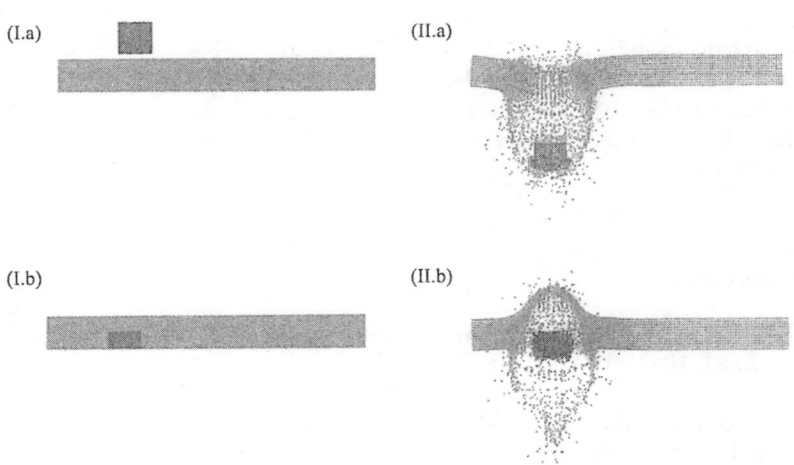

(I.a)   (II.a)

(I.b)   (II.b)

Figure 9 *Damage and fracture of tungsten cube and silica-phenolic bar*

5. Concluding remarks

It has been quoted in the literature that the tensile instability generally experienced in Eulerian based meshless techniques can be eliminated by using a Lagrangian kernel, *i.e.* by letting the kernel be a function of reference or initial configuration. This paper has discussed alternative Lagrangian SPH formulations to overcome the problem of such tensile instabilities. Two different formulations namely have been discussed at length. Remarkably, both Lagrangian expressions for the internal forces preserve linear and angular momentum given that in the case of corrected SPH the derivatives of the kernel at the reference configuration satisfy zero order completeness. As the derivatives of kernel functions are fixed at the reference configuration the amount of computational effort here would be much less than that of Eulerian based meshless technique. Further, for the alternative Lagrangian SPH formulation presented, as the corrections are already included in the definition of the gradient functions, the computational cost can be further reduced. However, it is important to note that for problems involving large distortions, a Lagrangian formulation may require frequent updates of the reference configurations. Robust procedures for carrying out such updates are currently under investigation.

Finally the ability of the Lagrangian formulations to simulate complex problems is demonstrated by several numerical examples. All the numerical simulations are carried out in three spatial dimensions using a total Lagrangian approach. The results obtained in the Taylor bar impact problem show close agreement with finite element simulations. The other simulations illustrate the possible potential applications of the Lagrangian SPH formulations.

Acknowledgements

This research has been funded by the Engineering and Physical Sciences Research Council (ESPRC) via grant number GR/L78352. This financial support is gratefully acknowledged.

6. References

1. L.B. Lucy, "A Numerical Approach to the Testing of the Fission Hypothesis", *Astro. J.*, 82, 1013, (1977).

2. R.A. Gingold and J. J. Monaghan, "Smooth Particle Hydrodynamics: Theory and Application to Non-Spherical Stars", *Mon. Not. R. Astron. Soc.*, 181, 375, (1977).

3. M. Schussler and D. Schmitt, Comments on Smoothed Particle Hydrodynamics, *Astron. Astrophys.*, 97, 373, (1981).

4. R.A. Gingold and J. J. Monaghan, "Kernel Estimates as a Basis for General Particle Methods in Hydrodynamics", *J. Comp. Phys.*, 46, 429, (1982).

5. J.J. Monaghan, "An Introduction to SPH", *Comput. Phys. Comm.*, 48, 89–96, (1988).

6. L.D. Libersky, A.G. Petschek, T.C. Carney, J.R. Hipp and F.A. Allahadi, "High Strain Lagrangian Hydrodynamics", *J. Comput. Phys.*, 109, 67, (1993).

7. W. Benz and E. Asphaug, "Simulations of Brittle Solids using Smooth Particle Hydrodynamics", *Comput. Phys. Comm.*, 87, 253, (1995).

8. G.R. Johnson, R.A. Stryk and S.R. Beissel, "SPH for High Velocity Impact Computations", *Comput. Methods Appl. Mech. Engrg.*, 139, 347, (1996).

9. W.K. Liu, S. Jun and Y.F. Zhang, "Reproducing Kernel Particle Methods", *Int. J. Num. Meth. Engrg.*, 20, 1081–1106, (1995).

10. J. Bonet and S. Kulasegaram, "Correction and Stabilization of Smooth Particle Hydrodynamics Methods with Applications in Metal Forming Simulations", *Int. J. Num. Meth. Engrg.*, 47, 1189–1214, (2000).

11. T. Belytschko, Y. Guo, W.K. Liu and S.P. Xiao, "A unified stability analysis of meshless methods", *Int. J. Num. Meth. Engrng.*, 48, 1359–1400, (2000).

12. J. Bonet and S. Kulasegaram, "Remarks on Tension Instability of Eulerian and Lagrangian Corrected Smooth Particle Hydrodynamics (CSPH) methods", *Int. J. Num. Mthds in Engrg.*, 52, 1203–1220, (2001).

13. J.W. Swegle, D.L. Hicks and S.W. Attaway, "Smooth Particle Hydrodynamics Stability Analysis", *J. Comp. Phys.* 116, (1995).

14. C.T. Dyka and R.P. Ingel, "An approach for Tension Instability in Smooth Particle Hydrodynamics(SPH)", Computers & Structures 57, (1995).

15. C.T. Dyka and R.P. Ingel, "Stress Points for Tension Instability in SPH", *Int. J. Num. Meth. Engrng.*, 40, 2325–2341, (1997).

16. J. Bonet and R.D. Wood, *Non-linear Continuum Mechanics for Finite Element Analysis*, Cambridge Uni. Press, Cambridge, UK (1997).

17. J. Bonet and A.J. Burton, "A Simple Average Nodal Pressure Tetrahedral Element for Incompressible and Nearly Incompressible Dynamic Explicit Application", *Comm. Numer. Meth. Eng.*, 14 (5), 437–449, (1998).

18. P.W. Randles, T.C. Carney, L.D. Libersky, J.D. Renick and A.G. Petschek, "Calculation of Oblique Impact and Fracture of Tungsten Cubes Using Smoothed Particle Hydrodynamics", *Int. J. Imp. Engng.*, 17, 661–672, (1995).

19. P.W. Randles and J.A. Nemes, "A Continuum Damage Model for Thick Composite Materials Subjected to High-Rate Dynamic Loading", *Mechanics of Materials*, 13(1), 1–13, (1992).

20. T. Belytschko and M.O. Neal, "Contact-Impact by the Pinball Algorithm with Penalty and Lagrangian Methods", *Int. J. Num. Mthds. in Engrg.*, 31, 547–572, (1991).

21. R. Vignjevic, J. Campbell and L. Libersky, "A Treatment of Zero-Energy Modes in the Smoothed Particle Hydrodynamics Method", *Comput. Methods. Appl. Mech. Engrg.*, 184, 67–85, (2000).

Chapter 4

An Extended Approach to Error Control in Experimental and Numerical Data Smoothing and Evaluation Using the Meshless FDM

Józef Krok
Section of Computer Method in Mechanics
Cracow University of Technology, Cracow, Poland

1. Introduction

This work addresses the validation of a new approach proposed to control error in smoothing/approximation of experimental and numerical data. On the base of *a posteriori* error analysis of data, we introduce an adaptive procedure of experimental data collection, and evaluation is presented.

One often has to transfer discrete data known at certain points to other points, for instance one may need *e.g.* a much clearer picture or require data smoothing. Sometimes one may also need additional data. How can this be done at minimal loss of accuracy? Is it possible to measure the degree of information loss and if so, how? Is it possible to recover, as a by-product, additional information on the data (regularity, smoothness) and locations of data points (guaranteeing the highest accuracy: distributions of data points density and function gradients are similar). Answers to the above questions are crucial for proper interpretation of experimental/numerical data. May one find a positive answer to these questions?

The present research is concentrated on further development of an approximation technique of physical/numerical data, based on the MWLS (*moving weighted least squares*) [7, 8, 9, 11] and *finite difference formulae* (FDM) and formulation of a new approach to experimental data measurement planning and implementation.

It includes:

- validation of postprocessing techniques for data approximation done in a discrete form and an iterative approach to additional enhancement of data at new (required) locations,

- formulation of *a posteriori* error technique to trace loss of accuracy of original data by using different "error norms", *a posteriori* error estimation,
- evaluation of experimental points density in experimental data taking into account equal error distribution,
- formulation of the new adaptive approach to experiment planning and implementation, taking into account *a posteriori* error estimation and distribution of experimental points with equidistributed error,
- analysis of the wheel saw cut data, especially for the wheel #2 (see R. Czarnek [3], and J. Krok, J. Orkisz, A. Skrzat [11], J. Krok [8]), as an sample application of the proposed approach.

Part of the theoretical considerations is based on *adaptive finite element analysis* (AFEM). AFEM provides tools to solve the problem under consideration, even though the problem does not necessarily conform to the AFEM case, because several assumptions are violated (for example one does not know the rate of convergence and degree of smoothness, *i.e.* regularity of the physical data). In this work, we develop an original idea to interpret the physical data in the same way as the numerical data coming from FEM or FDM – using an adaptive analysis.

2. Approximation and error analysis of numerical and experimental data

The theory of *a posteriori* error estimation in discrete methods like in the FEM or MFDM is already well established. As a result one obtains new mesh density to solve boundary value problems with highest possible accuracy *i.e.* with equidistributed errors. Now the same idea is proposed for experimental mechanics. Theory presented here allows one to evaluate results obtained in experiment and to give very precise information on location and density of gauges or on the size of a moire interferometry grid (output of any experimental method may be evaluated). If it is not possible to improve measurement quality, one gets precise information on data measured with insufficient precision. Reliability indices defined in present work permit one to take into account measurements differing in quality.

A posteriori error estimation in FEM, which may be (however indirectly) used to introduce the proposed idea, is presented here to explain our intent.

2.1. *Zienkiewicz–Zhu* **a posteriori** *error estimator*

First, for the sake of clarity of proposed experimental data approximation, the *a posteriori* error estimator used in FEM and MFDM, based on the postprocessing approach will be presented. For estimators based on the stress (or flux) recovery technique (Zienkiewicz, Zhu [14–17]) one has

$$\|e\| = \left[\int_{\Omega} (\sigma - \sigma^h)^T D^{-1} (\sigma - \sigma^h) d\Omega \right]^{1/2} \tag{1}$$

where σ^h are stresses obtained by the FEM.

The exact stresses σ are approximated by new stresses obtained using the stress recovery procedure (the meshless finite difference method is used here [9, 15])

$$\sigma^* = N\overline{\sigma}^* \tag{2}$$

where $\overline{\sigma}^*$ are nodal values obtained by the MFDM recovery procedure, and N is a shape functions matrix.

The exact solution for the strain energy is estimated as

$$\|U\| = \left[\int_{\Omega} (\sigma^h)^T D^{-1} (\sigma^h) d\Omega \right]^{1/2} + \|e\|^2 \tag{3}$$

where an error of the energy norm is expressed as

$$\|e\| = \left[\int_{\Omega} (\sigma^* - \sigma^h)^T D^{-1} (\sigma^* - \sigma^h) d\Omega \right]^{1/2} \tag{4}$$

Both $\|e\|$ and $\|U\|$ norms may be evaluated as a sum of their respective element contributions so that (n denotes the total number of elements in the mesh)

$$\|e\|^2 = \sum_{i=1}^{n} \|e\|_i^2, \quad \|U\|^2 = \sum_{i=1}^{n} \|U\|_i^2 \tag{5}$$

Remark. *An a posteriori error procedure can be split into two main stages:*

- *stage 1: calculation of stresses (or other primary values) at Gaussian points – the primary set of points,*
- *stage 2: approximation of the Gaussian-located stresses at nodal positions (secondary set of points), retrieval of the nodal values to Gauss points using (for example) standard shape functions or other kind of approximation.*

Having two sets of values of different accuracy at the same points, one may calculate local $\|e\|_i$, $\|U\|_i$ and global $\|e\|$, $\|U\|$ norms.

2.2. Meshless error estimator

A new idea of *a posteriori* error estimation of randomly distributed experimental data or numerical data coming from FEM or MFDM analysis is presented here.

Let us define the following problem:

- data, (not necessary stresses like in equation (4)) coming from experiment, located at certain points – set #1 (see experimental points – set of primary points – Figure 1) is given,
- the fictitious sets of points used later in calculation – set #2 (see fictitious points – set of secondary points – Figure 1) is given.

The problem lies in data translation (approximation) from #1 points set to #2 points set. The problem is exactly the same as in the error estimator (4), but now one has two different sets of points with, in fact, arbitrary (not elemental) locations and one has no information on regularity, smoothness and reliability of the data.

Differences between two surfaces defined by data #1 and #2 may be measured by the norm

$$\|e\| = \left[\int_\Omega (u^* - u^h)^T (u^* - u^h) + (\nabla u^* - \nabla u^h)^T (\nabla u^* - \nabla u^h) \right.$$

$$\left. + (\kappa u^* - \kappa u^h)^T (\kappa u^* - \kappa u^h) d\Omega \right]^{1/2} \tag{6}$$

where u^h is the vector of experimental data (in experimental points) and u^* is the vector of fictitious sought data. Sometimes weighting factors may be used to equilibrate dimensions of terms. In the above formula one can omit (sometimes not) the gradients and curvature terms (κ – see generalized curvature – Karmowski and Orkisz [5, 6]). One can also use a discrete form of this formula, summing up differences between values at experimental points.

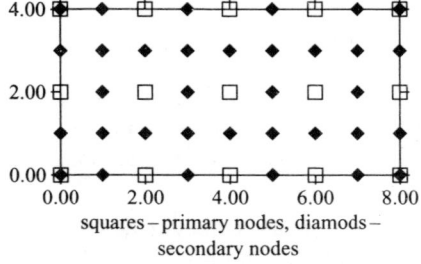

squares – primary nodes, diamods –
secondary nodes

Figure 1 *Primary and secondary mesh for approximation of physical or numerical data*

To solve this problem, data from experimental points is approximated to fictitious ones (using FDM approximation – see next part of this work) and later on, taking values at fictitious points as original data, approximated back from fictitious points to experimental ones. In this two-stage approximation part of data is lost, but if differences between original data in the experimental points and fictitious data in the same points are small enough, one may expect that the approximation in the first step does not introduce too large error. As will be seen from numerical analysis this assumption is true.

Additionally, in the zones where the gradients of approximated function are larger, the error magnitude is considerably higher as compared with the zones with smaller data gradients. Moreover, if irregularity in data is large the error increases. Those facts may be used, as by-product important information, to evaluate experimental data. Having the vector of differences between experimental and fictitious values at experimental points one can "smear" the error, approximating the vector of differences from experimental to fictitious points. Adding correction to initial fictitious values one can obtain new enhanced fictitious values. This process can be repeated (iteration process gives possibility to avoid fluctuation, especially when data is very smooth, like MFDM solution). In this way, the usual approach elaborated mainly in an adaptive finite element method is unified, extended and generalized.

The total norm of the measured values may be expressed as (the discrete form of the below norms may be used):

$$\|U\|^2 = \left[\int_\Omega (u^h)^T u^h + (\nabla u^h)^T \nabla u^h + (\kappa u^h)^T \kappa u^h d\Omega \right] + \|e\|^2 \qquad (7)$$

The key question is whether one can evaluate experimental data using norms (6) and (7)? The answer is 'yes', if data are regular enough.

As one can see from equation (7), not only values of a function measured, but gradients of the function and curvatures are taken into account as well. Additionally, one can considerably improve the coarse FEM solution, using MFDM approximation to find any required derivatives at any point of domain, with error control as a by-product.

3. Meshless finite difference approximation

The approximation $u^h(\mathbf{x})$ of function $u(\mathbf{x})$ is posed as polynomial of order m with non-constant coefficients $a_0(\mathbf{x}), a_1(\mathbf{x}), \ldots, a_m(\mathbf{x})$. The order of polynomial is defined as the order of the basis. For a linear basis in two dimensions $u^h(\mathbf{x})$ can be written as

$$u^h(\mathbf{x}) = a_0 + a_1 x + a_2 y, \qquad (8)$$

where unknown parameters $a_j(\mathbf{x})$ vary with $\mathbf{x}$. The local approximation (for $\bar{\mathbf{x}} = \mathbf{x}$) is given by [1, 9, 10, 15]

$$u^h(\mathbf{x}, \bar{\mathbf{x}}) = \sum_{j=0}^{m} p(\mathbf{x}) a_j(\bar{\mathbf{x}}) = \mathbf{p}^T(\mathbf{x})\mathbf{a}(\mathbf{x}) \tag{9}$$

where $\mathbf{p}(\mathbf{x})$ is a complete polynomial of order m

$$\mathbf{p}^T(\mathbf{x}) = [1, x, y, x^2, xy, y^2, \ldots,] \tag{10}$$

and $\mathbf{a}(\mathbf{x})$ is given by

$$\mathbf{a}^T(\mathbf{x}) = [a_0(\mathbf{x}), a_1(\mathbf{x}), \ldots, a_m(\mathbf{x})]. \tag{11}$$

The unknown parameters $a_j(\mathbf{x})$ at any given point are determined by minimizing the difference between the local approximation at that point and the nodal parameters u_I i.e. weighted, discrete L_2 norm

$$J(\mathbf{a}) = \sum_{I=1}^{n} w(\mathbf{x} - \mathbf{x}_I)[u^h(\mathbf{x}_I, \mathbf{x}) - u_I]^2 = \sum_{I=1}^{n} w(\mathbf{x} - \mathbf{x}_I)[p^T(\mathbf{x}_I)a(\mathbf{x}) - u_I]^2, \tag{12}$$

where $w(\mathbf{x} - \mathbf{x}_I)$ is a shift of a given weighting function $w(\mathbf{x})$, and n is the number of nodes in the neighborhood of $\mathbf{x}$ for which the weighting function $w(\mathbf{x} - \mathbf{x}_I) \neq 0$.

The minimum of J in (12) with respect to $\mathbf{a}(\mathbf{x})$ leads to the following set of linear equation

$$\mathbf{A}(\mathbf{x})\mathbf{a}(\mathbf{x}) = \mathbf{B}(\mathbf{x})\mathbf{u}. \tag{13}$$

After solving the set of equation (13), one obtains

$$\mathbf{a}(\mathbf{x}) = \mathbf{A}^{-1}(\mathbf{x})\mathbf{B}(\mathbf{x})\mathbf{u} = \sum_{I=1}^{n} \mathbf{A}^{-1}(\mathbf{x})\mathbf{B}_I(\mathbf{x})u_I = \mathbf{Q}(\mathbf{x})\mathbf{u}, \tag{14}$$

where

$$\mathbf{A}(\mathbf{x}) = \sum_{I=1}^{n} w(\mathbf{x} - \mathbf{x}_I)\mathbf{p}(\mathbf{x}_I)\mathbf{p}^T(\mathbf{x}_I), \tag{15}$$

$$\mathbf{B}(\mathbf{x}) = [w(\mathbf{x} - \mathbf{x}_1)\mathbf{p}(\mathbf{x}_1), w(\mathbf{x} - \mathbf{x}_2)\mathbf{p}(\mathbf{x}_2), \ldots, w(\mathbf{x} - \mathbf{x}_n)\mathbf{p}(\mathbf{x}_n)]. \tag{16}$$

Substituting (14) into (9), the MWLS approximates can be defined as

$$u^h = \mathbf{p}^T(\mathbf{x})\mathbf{A}^{-1}(\mathbf{x})\mathbf{B}(\mathbf{x})\mathbf{u}$$

$$= \sum_{I=1}^{n}\sum_{j=0}^{m} p_j(\mathbf{x})[\mathbf{A}^{-1}(\mathbf{x})\mathbf{B}(\mathbf{x})]_{jI}u_I = \sum_{I=1}^{n}\tilde{N}_I(\mathbf{x})u_I = \tilde{N}\mathbf{u}, \qquad (17)$$

where

$$\tilde{N}_I(\mathbf{x}) = \sum_{j=0}^{m} p_j(\mathbf{x})[\mathbf{A}^{-1}(\mathbf{x})\mathbf{B}(\mathbf{x})]_{jI} = \mathbf{p}^T\mathbf{A}^{-1}\mathbf{B}_I = \sum_{j=1}^{m} p_j Q_{jI} \qquad (18)$$

are the shape functions in MWLS approximation (note that $\sum^n \tilde{N}_I = 1$).

To determine the derivatives of the approximating function $u^h(\mathbf{x})$, one has to obtain the shape functions' derivatives. The spatial derivatives of the shape functions are determined by

$$\tilde{N}_{I,x} = [\mathbf{p}^T\mathbf{A}^{-1}\mathbf{B}_I]_{,x} = \mathbf{p}^T_{,x}\mathbf{A}^{-1}\mathbf{B}_I + \mathbf{p}^T\mathbf{A}^{-1}_{,x}\mathbf{B}_I + \mathbf{p}^T\mathbf{A}^{-1}\mathbf{B}_{I,x}, \qquad (19)$$

where

$$\mathbf{B}_I(\mathbf{x}) = \frac{\partial w(\mathbf{x}-\mathbf{x}_I)}{\partial x}\mathbf{p}(\mathbf{x}_I). \qquad (20)$$

Matrix $\mathbf{A}^{-1}_{,x}$ is computed by

$$\mathbf{A}^{-1}_{,x} = -\mathbf{A}^{-1}\mathbf{A}_{,x}\mathbf{A}^{-1} \qquad (21)$$

where

$$A_{,x} = \sum_{I=1}^{n} \frac{\partial w(\mathbf{x}-\mathbf{x}_I)}{\partial x}\mathbf{p}(\mathbf{x}_I)\mathbf{p}^T(\mathbf{x}_I). \qquad (22)$$

To compute the shape functions and their derivatives, the $\mathbf{A}$ matrix has to be inverted. This process is more computationally efficient if LU decomposition of the matrix $\mathbf{A}$ is performed. The shape functions in (18) can be written as

$$\tilde{N}_I = \sum_{j=0}^{m} p_j(\mathbf{x})\mathbf{A}^{-1}(\mathbf{x})\mathbf{B}_{jI}(\mathbf{x}) = \mathbf{p}^T\mathbf{A}^{-1}\mathbf{B}_I = \mathbf{g}^T\mathbf{B}_I, \qquad (23)$$

where the following relationship was used [1]

$$\mathbf{A}(\mathbf{x})\mathbf{g}(\mathbf{x}) = \mathbf{p}(\mathbf{x}). \qquad (24)$$

The vector $\mathbf{g}(\mathbf{x})$ can be determined the same way as the $\mathbf{a}$ vector. The derivatives of vector $\mathbf{g}(\mathbf{x})$ can be computed similarly; this leads to a computationally efficient procedure to determine derivatives of u^h. Taking spatial derivatives of (24) and rearranging these, one has

$$\mathbf{A}(\mathbf{x})\mathbf{g}(\mathbf{x}),_x = \mathbf{p}(\mathbf{x}),_x - \mathbf{A},_x\mathbf{g}. \tag{25}$$

Thus, the derivative of $\mathbf{g}(\mathbf{x})$ can be calculated using the same **LU** decomposition obtained from (24). Spatial derivatives of the shape function may be obtained as [1]

$$\tilde{N}_I(\mathbf{x}),_x = \mathbf{g}(\mathbf{x}),_x\mathbf{B}_I + \mathbf{g}(\mathbf{x})\mathbf{B}_{I,x}. \tag{26}$$

By consecutive derivation of equation (25) one obtains the set of following equations for vector $\mathbf{g}$ and its derivatives (once more note that the **LU** decomposition of **A** matrix is performed only once):

$$\begin{aligned}
\mathbf{A}(\mathbf{x})\mathbf{g}(\mathbf{x}) &= \mathbf{p}(\mathbf{x}), \\
\mathbf{A}(\mathbf{x})\mathbf{g}(\mathbf{x}),_x &= \mathbf{p}(\mathbf{x}),_x - \mathbf{A},_x\mathbf{g}, \\
\mathbf{A}(\mathbf{x})\mathbf{g}(\mathbf{x}),_y &= \mathbf{p}(\mathbf{x}),_y - \mathbf{A},_y\mathbf{g}, \\
\mathbf{A}(\mathbf{x})\mathbf{g}(\mathbf{x}),_{xx} &= \mathbf{p}(\mathbf{x}),_{xx} - \mathbf{A},_{xx}\mathbf{g} - 2\mathbf{A},_x\mathbf{g},_x, \\
\mathbf{A}(\mathbf{x})\mathbf{g}(\mathbf{x}),_{xy} &= \mathbf{p}(\mathbf{x}),_{xy} - \mathbf{A},_{xy}\mathbf{g} - \mathbf{A},_x\mathbf{g},_y - \mathbf{A},_y\mathbf{g},_x, \\
\mathbf{A}(\mathbf{x})\mathbf{g}(\mathbf{x}),_{yy} &= \mathbf{p}(\mathbf{x}),_{yy} - \mathbf{A},_{yy}\mathbf{g} - 2\mathbf{A},_y\mathbf{g},_y.
\end{aligned} \tag{27}$$

This leads to a simple relationship for the derivatives of the shape functions up to the second order

$$\begin{aligned}
\tilde{N}_I &= \mathbf{g}(\mathbf{x})\mathbf{B}_I, \\
\tilde{N}_{I,x} &= \mathbf{g}(\mathbf{x}),_x\mathbf{B}_I + \mathbf{g}(\mathbf{x})\mathbf{B}_{I,x}, \\
\tilde{N}_{I,y} &= \mathbf{g}(\mathbf{x}),_y\mathbf{B}_I + \mathbf{g}(\mathbf{x})\mathbf{B}_{I,y}, \\
\tilde{N}_{I,xx} &= \mathbf{g}(\mathbf{x}),_{xx}\mathbf{B}_I + 2\mathbf{g}(\mathbf{x}),_x\mathbf{B}_{I,x} + \mathbf{g}(\mathbf{x})\mathbf{B}_{I,xx}, \\
\tilde{N}_{I,xy} &= \mathbf{g}(\mathbf{x}),_{xy}\mathbf{B}_I + \mathbf{g}(\mathbf{x}),_x\mathbf{B}_{I,y} + \mathbf{g}(\mathbf{x}),_y\mathbf{B}_{I,x} + \mathbf{g}(\mathbf{x})\mathbf{B}_{I,xy}, \\
\tilde{N}_{I,yy} &= \mathbf{g}(\mathbf{x}),_{yy}\mathbf{B}_I + 2\mathbf{g}(\mathbf{x}),_y\mathbf{B}_{I,y} + \mathbf{g}(\mathbf{x})\mathbf{B}_{I,yy}.
\end{aligned} \tag{28}$$

In practical calculations, the local coordinate system $\mathbf{h} = \mathbf{h}(h, k)$ is used

$$\mathbf{h} = \mathbf{h}(h, k) = \mathbf{x} - \mathbf{x}_0, \quad \mathbf{x}_0 = (x_0, y_0), \tag{29}$$

where $\mathbf{x}_0 = \mathbf{x}_0(x_0, y_0)$ is the point in which approximation is sought. The base vector is taken as

$$p^T = \left[1, h, k, \frac{1}{2}h^2, hk, \frac{1}{2}k^2, \ldots \right], \tag{30}$$

so coefficients $a_0, a_1, \ldots, a_m$ may be immediately interpreted as derivatives at point x_0 (usually called local derivatives)

$$a_0 = u_0, a_1 = \frac{\partial u_0}{\partial x}, a_2 = \frac{\partial u_0}{\partial y}, \ldots, \tag{31}$$

and the matrix $\mathbf{Q}$ (see (14)) now is a generalized FD matrix. Combination of rows of this matrix and a vector of nodal values yields immediately values of function and its derivatives at point x_0 (but these derivatives may not be continuous from point to point).

Consistent meshless finite difference matrix $\tilde{\mathbf{Q}}$ and the approximation rule now have the form

$$\mathbf{D}u = \begin{Bmatrix} u \\ u_x \\ u_y \\ u_{xx} \\ u_{xy} \\ u_{yy} \end{Bmatrix} = \begin{bmatrix} \tilde{N}_1 & \tilde{N}_2 & \cdots & \tilde{N}_n \\ \partial \tilde{N}_1/\partial x & \partial \tilde{N}_2/\partial x & \cdots & \partial \tilde{N}_n/\partial x \\ \vdots & \vdots & \vdots & \vdots \\ \partial^2 \tilde{N}_1/\partial y^2 & \partial^2 \tilde{N}_2/\partial y^2 & \cdots & \partial^2 \tilde{N}_n/\partial y^2 \end{bmatrix} \begin{Bmatrix} u_1 \\ u_2 \\ \vdots \\ u_n \end{Bmatrix} = \tilde{\mathbf{Q}}u \tag{32}$$

The explicit form of the matrix $\tilde{\mathbf{Q}}$ has the following form (note that matrix $\mathbf{Q}_I$ contains columns of the approximation matrix $\mathbf{Q}$ of zero-th order)

$$\tilde{\mathbf{Q}} = \begin{bmatrix} \mathbf{p}^T \mathbf{Q}_1 & \cdots & \mathbf{p}^T \mathbf{Q}_n \\ \mathbf{p}_{,x}^T \mathbf{Q}_1 + \mathbf{p}^T \mathbf{Q}_{1,x} & \cdots & \mathbf{p}_{,x}^T \mathbf{Q}_n + \mathbf{p}^T \mathbf{Q}_{n,x} \\ \mathbf{p}_{,y}^T \mathbf{Q}_1 + \mathbf{p}^T \mathbf{Q}_{1,y} & \cdots & \mathbf{p}_{,y}^T \mathbf{Q}_n + \mathbf{p}^T \mathbf{Q}_{n,y} \\ \mathbf{p}_{,xx}^T \mathbf{Q}_1 + 2\mathbf{p}_{,x}^T \mathbf{Q}_{1,x} + \mathbf{p}^T \mathbf{Q}_{1,xx} & \cdots & \mathbf{p}_{,xx}^T \mathbf{Q}_n + 2\mathbf{p}_{,x}^T \mathbf{Q}_{n,x} + \mathbf{p}^T \mathbf{Q}_{n,xx} \\ \mathbf{p}_{,xy}^T \mathbf{Q}_1 + \mathbf{p}_{,x}^T \mathbf{Q}_{1,y} + \mathbf{p}_{,y}^T \mathbf{Q}_{1,x} + \mathbf{p}^T \mathbf{Q}_{1,xy} & \cdots & \mathbf{p}_{,xy}^T \mathbf{Q}_n + \mathbf{p}_{,x}^T \mathbf{Q}_{n,y} + \mathbf{p}_{,y}^T \mathbf{Q}_{n,x} + \mathbf{p}^T \mathbf{Q}_{n,xy} \\ \mathbf{p}_{,yy}^T \mathbf{Q}_1 + 2\mathbf{p}_{,y}^T \mathbf{Q}_{1,y} + \mathbf{p}^T \mathbf{Q}_{1,yy} & \cdots & \mathbf{p}_{,yy}^T \mathbf{Q}_n + 2\mathbf{p}_{,y}^T \mathbf{Q}_{n,y} + \mathbf{p}^T \mathbf{Q}_{n,yy} \end{bmatrix} \tag{33}$$

Taking into account that approximation is sought in the origin of local coordinate system one has $\mathbf{p}^T(\mathbf{0}) = [1, 0, 0, 0, 0, 0, \ldots]$, and thus (note: Q_{1i} is the first element in each i-th column $\mathbf{Q}_i$ of FD matrix)

$$\tilde{N}_i = \mathbf{p}^T \mathbf{Q}_i = Q_{1i}, \tag{34}$$

As one may see from equation (34) *global* shape functions are equal to *first row elements of FD matrix.*

This means that global approximation is exactly the same as a local one in origin of a local coordinate system. The meshless shape functions and the diffuse derivatives result naturally from the moving least squares formulation introduced in the scope of computational mechanics [13, 14]. Further bibliography and historical details may be found in the reference books [19].

Relation (32) is valid for a very wide class of meshless approaches and yields continuous derivatives up to the second order very easily. Having two different approximation matrices: $\mathbf{Q}$ – obtained from the MWLS approximation (see equation (14)) and $\tilde{\mathbf{Q}}$ – (32) – obtained by means of the direct differentiation of $u^h(\mathbf{x})$, one has another, very useful, capability to measure error as a violation of continuity in approximation of the first and second derivatives. This way one may have at the same time two different approximation matrices.

The continuity feature of derivatives is not always beneficial, especially when approximation of data given in a set of arbitrarily spaced points is needed. Besides, there are problems with proper definition of a weighting function on arbitrarily spaced grid of points, because results of approximation, especially derivative values, strongly depend on the type of weighting functions used (dimensions of weight support). If support of the weighting function is not properly correlated with grid density and its form is not appropriate for the purpose required, results may be considerably worse than in the case of direct MWLS MFDM approximation. If support of the approximant is too large, approximation is too smooth and thus local peak values of approximated function are anihilated.

The weighting function used here is [5, 6]

$$w(\rho) = \left(\rho^2 + \frac{g^4}{g^2 + \rho^2}\right)^{-(p+1)}, \tag{35}$$

where ρ is the distance between the central point and the node, p denotes polynomial order and g is an optimality parameter making singular weighting function (interpolation) or non-singular weights (approximation) available. If the optimality parameter g tends to a small value, the weighting function enforces interpolation. If the optimality parameter tends to large number, approximation takes place, but data smoothing may be over emphasized.

It is worth mentioning that the continuity problem arises in the MWLS approximation. Continuity requires that either all nodes in the domain considered are taken into account each time or weighting functions defined on an appropriate finite supports are used providing zero end conditions. If such support is not properly corelated with the mesh density, approximation results may be of considerably lower quality than they could be. On the other hand continuity features of MWLS approximation and its derivatives may be not needed in practice (see [12, 13]).

Test problem [12]

Though the matter requires a deeper and systematic study (see [12]) a valuable insight into the MWLS approximation quality was gained by analysis of a simple test.

Considered was a set of data presenting the values of function $u = \sqrt{25 - x^2}$ defined at nodes of an evenly spaced mesh: 0, 0.5, 1.0, 1.5, 2.0, 2.5, 3.0 having the increment 0.5.

The local (31) and global (17) MWLS approximation was performed for the function itself and for its first local (31) and global (19) derivatives. Results of error analysis obtained in the interval [0.5, 3.0] for various weighting functions are presented and compared in Figures 2a and 2b. The following may then be noticed:

– Results of the local (3rd order) and global (consistent) approximation are, of course, the same for the function itself when using the same weighting functions.
– Neither method showed clear advantages with respect to result quality when comparing the first derivative found by means of either the local (31) or the global (consistent) (19) differentiation approach. For local derivatives superconvergence property at internal nodes is noted (error of the local derivatives is considerably lower than error of the consistent derivatives). Superconvergence property of local derivatives for lower approximation order is much stronger than for higher order (not presented here). On the other hand approximation error of the consistent derivatives is more uniform. Maximal error is lower than approximation error of the local derivatives. It is interesting to see that the gap between local and global derivatives is proportional to

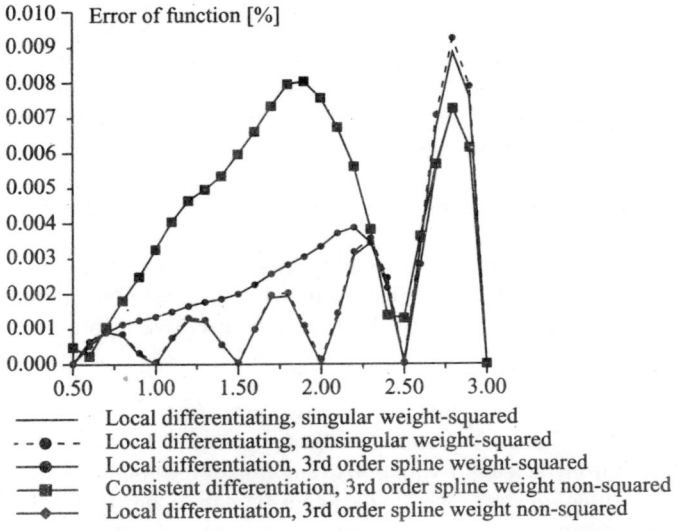

Figure 2a *Weight function influence on results of approximation – function error data sought at points: min = 0.5, max = 3.0, increment $\Delta x = 0.1$*

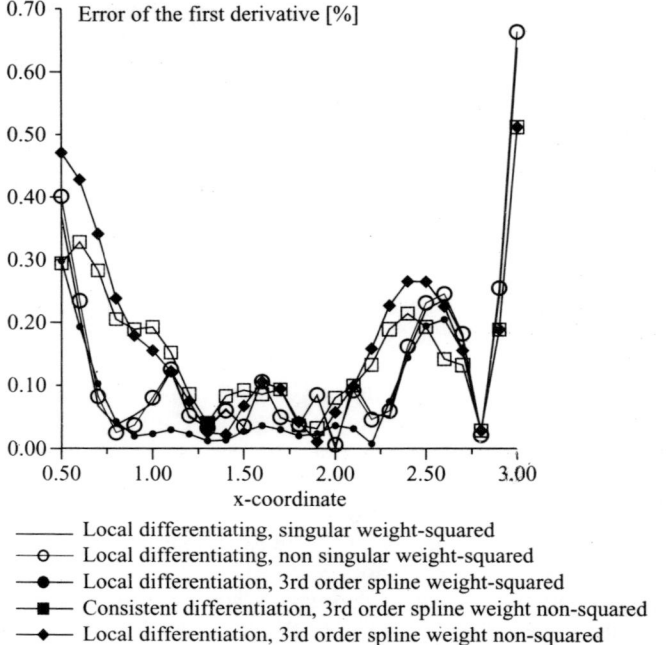

Figure 2b *Weight function influence on results of approximation-derivative error*

the approximation error. This fact explains why the error estimator proposed by Gavette, Cuesta and Ruiz works very well [18]. Probably, for the first time, it is possible to define a very convenient error estimator in mesh-less methods, based on postprocessing, but for the two different types of derivatives.

– Squared weighting functions proved clear advantage (minimal errors) over non-squared ones.

– The smallest errors in the function approximation were observed when the singular weighting factor (33) was used , while squared non-singular 3rd order spline weight [1], was found the best for derivative evaluation.

From above test it is evident that there are two problems: 1^0 discrete data approximation problem, 2^0 boundary-value solution problem. It is not justified to extend the conclusion from data approximation to solution of the boundary value problem. Even if approximation works well in data approximation, one may not obtain good results when a boundary-value problem solution is needed. On the other hand, bad results of data approximation do not necessarily mean that approximation will give bad results during solution of the boundary-value problem.

4. Definition of the points density function in experimental and numerical discrete data

4.1. *Acceptable solution and mesh (grid) refinements function*

In an adaptive solution approach, the *a posteriori* errors are used for appropriate mesh modifications by means of the so-called error indicators and mesh refinement parameters. An approach to mesh modification applied to the adaptive FEM (or validation of a density of experimental points) is discussed below. This problem is very important because one has to have the capability to take into account, in numerical as well as in physical experiments, relation between gradients of the measured function and density (location) of the points at which information is available.

The solution is 'correct' if the two following conditions are satisfied:

(i) The global error in the energy norm is less than a specified percentage value of the total strain energy

$$\|e\| = \eta\|U\| \tag{36}$$

where η is the 'USER' specified value of a permissible relative global error.

(ii) Distribution of elements in a new mesh satisfies a local mesh optimality criterion

$$\|e\|_i = \|e\|_{all(i)} \tag{37}$$

where $\|e\|_i$ is the actual error norm in i-th element and $\|e\|_{all(i)}$ is the 'required' error norm in the element.

The global and local error parameters may be defined from equations (36) and (37) as

$$\xi_g = \frac{\|e\|}{\eta\|U\|}, \quad \bar{\xi}_i = \frac{\|e\|_i}{\|e\|_{all(i)}}. \tag{38}$$

The mesh refinement parameter for the i-th element is introduced as a combination of the global and local parameters [2]

$$\xi_i = \bar{\xi}_i \xi_g = \frac{\|e\|\|e\|_i}{\eta\|U\|\|e\|_{all(i)}}. \tag{39}$$

One of the most important questions is: how one can define the required error norm for each element. The following definitions are considered here:

(i) the global error, equally distributed all over elements in the mesh (Zienkiewicz–Zhu [15])

$$\|e\|_{all(i)} = \frac{\|e\|}{\sqrt{n}}, \tag{40}$$

where n is the total number of elements in a mesh.

(ii)   mesh is optimal if squared error per unit element volume is the same over the whole mesh *i.e.* (Bugeda, Onate [2]); taking also into account equation (37) one has

$$\frac{\|e\|_i}{(\Omega_i)^{1/2}} = \frac{\|e\|}{(\Omega)^{1/2}} \quad \text{and} \quad \|e\|_i = \|e\|_{all(i)} = \|e\| \left(\frac{\Omega_i}{\Omega} \right)^{1/2} \quad (41)$$

Using equations (40) and (41) one may obtain the following *element refinement parameters*

$$\xi_i = \frac{\|e\|_i}{\eta \|U\| (n)^{-1/2}}, \quad \xi_i = \frac{\|e\|_i}{\eta \|U\|} \left(\frac{\Omega}{\Omega_i} \right)^{1/2} \quad (42)$$

for equal error distribution [15] and for the equal specific error distribution [2]. However, one should notice that the element and global error norms have different orders of convergence

$$\|e\|_i \approx O(h_i^m) \Omega^{1/2} \approx O \left(h_i^{m+d/2} \right), \quad \|e\| \approx O(h^m) \quad (43)$$

where h_i and h are the i-th element size and average size of all the elements in the mesh, m is the element order and d is the problem dimension. Dividing the element error by its area one obtains

$$\frac{\|e\|_i}{(\Omega_i)^{1/2}} \approx O(h_i^m) \quad (44)$$

and new element size parameter may be defined as (this is valid only for FEM)

$$\xi_{new} = (\overline{\xi_i} \xi_g)^{1/m} = (\xi_i)^{1/m}. \quad (45)$$

Refinement of the mesh may be carried out in two completely different ways: breaking elements (so called constrained approximation – see Oden, Demkowicz *et al.* [4]), or remeshing (Zhu, Zienkiewicz [15]). For the purpose of this research, th remeshing technique is preferred, as it is compatible with both FEM and MFDM discretizations.

4.2. *Error and grid density evaluation in saw cut experiment*

Crack nucleation propagation and failure of railroad car wheels is greatly influenced by residual stresses existing in those wheels, as a result of manufacturing and service conditions. The knowledge of residual stress distribution in wheels is thus required. Experimental data used for residual stress reconstruction is collected during radial saw cutting of a wheel in laboratory conditions in order to relieve residual stresses and strains, see Figure 3. In order to obtain reasonable residual stress estimation,

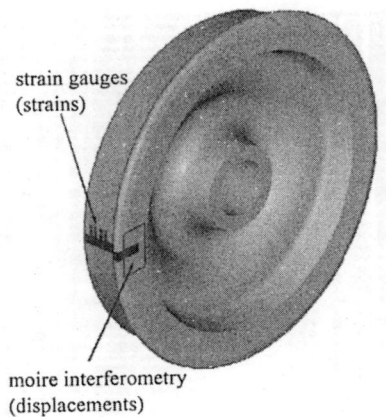

strain gauges
(strains)

moire interferometry
(displacements)

Figure 3 *Measurements taken at saw cut [19]*

the additional approximation process is applied, which simultaneously uses error estimation procedures for the problem considered.

Wheel saw cut experimental data may be evaluated using equations (36) and (37). Of course, density of experimental points depends on local (37) conditions, i.e. error at experimental points must be bound by a certain admissible value. *The global and local error refinement parameters* may be defined from equations (36) and (37) as in FEM analysis. Combining both global (36) and local (37) criteria one obtains the same formula *for the mesh refinement parameter* at the i-th experimental point like in the FEM. What does mesh refinement parameter in experiment mean? It means that an experimental value at certain points changes too rapidly when compared with the mean value and local density of experimental points. In other words, density of experimental points must be increased in certain part of the region, it is simply to low to properly describe the gradients of the measured function. New, required density is computed by formula $(42)_1$ – discrete form or $(42)_2$ – continuous form. One can take into account a weighting factor like an area assigned to an experimental point (see equation (42)). This is a proper definition of the admissible error at a point. Equations (43), (44) and (45) are not valid here because one does not have any information on the convergence of experimental results with respect to the density of experimental points. This very important problem is still open.

5. Approximation of the physical data and preliminary *a posteriori* error analysis by MFDM formulas

An error analysis described above has been applied to a problem of wheel saw cut data approximation and to calculate influence matrix coefficients (see [11]). The meshless finite difference approximation [1, 8, 9, 11] has been applied. In

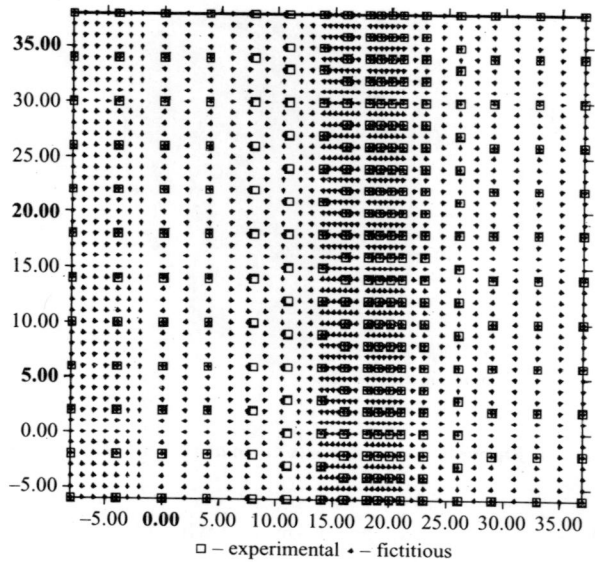

□ – experimental • – fictitious

Figure 4 *Flange side of the wheel, experimental and fictitious grids*

the examples presented fictitious (discrete!) mesh generated previously: Figure 4 (flange side of the wheel) has been used and the error has been determined at the experimental points.

In numerical calculations the physical data coming from the wheel #2 cutting process have been considered. Twenty different sets of data: horizontal (circumferential) and vertical (radial) displacements coming from five cuts (see R. Czarnek [3] and J. Krok [8]) of the wheel have been used in calculations.

Three different effects concerning experimental data are investigated:

1. An approximation error of the measured values from experimental grid to one used in numerical analysis.
2. Evaluation of the measured values taking into account five different 'error' norms.
3. Estimation of the new experimental points' grid density with equal distribution of an approximation error kept in mind.

In both analyses, local as well as global error norms were considered.

An approximation of experimental data with different order and different number of nodes in stars is presented in Figures 5–28. Notation used: n_taylor – number of coefficients in Taylor series expansion, nodes – number of nodes in star,

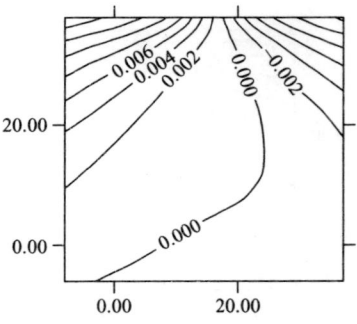

Figure 5 *Cut #1, flange side, horizontal displacements, original data, min = −1.25E−2, max = 1.55E−2*

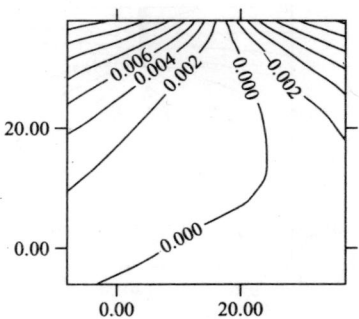

Figure 6 *Flange side, horizontal displacements, approximated data after 7 iterations*

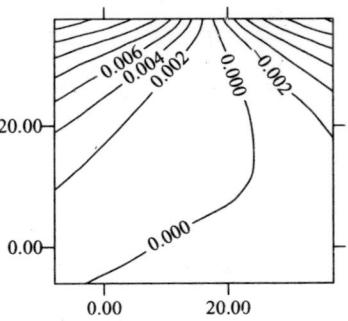

Figure 7 *Flange side, horizontal displacements, recovery data, no iteration*

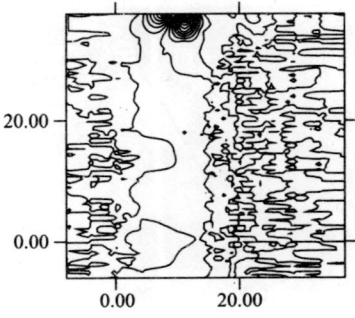

Figure 8 *Flange side, error of the horizontal displacements, no iterations, min = 1.35E−5, max = 7.46E−6, inc = 1E−6*

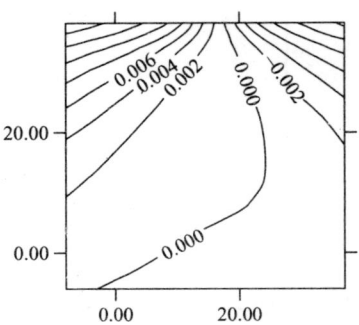

Figure 9 *Flange side, horizontal displacements, recovery data, after iterations*

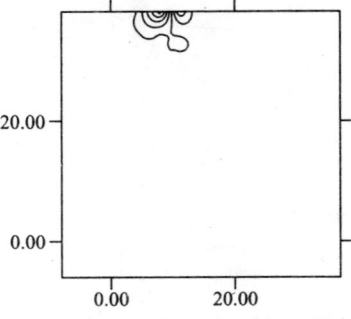

Figure 10 *Flange side, error of the horizontal displacements, after 7 iterations, min = −1.34E−6, max = 2.40E−6, inc = 5E−7*

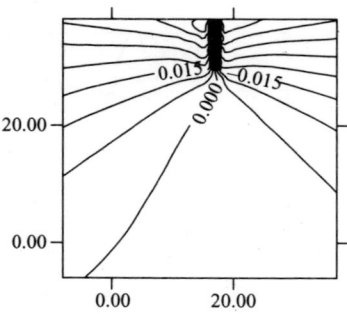

Figure 11 *Cut #3, flange side, horizontal displacements, original data, min* $= -4.48E-2$, *max* $= 4.33E-2$

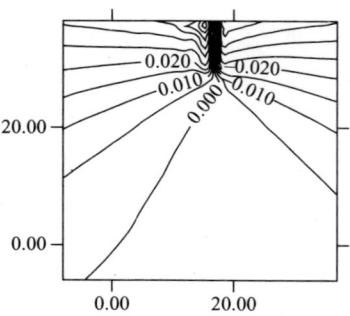

Figure 12 *Flange side, horizontal displacements, approximated data, after 7 iterations*

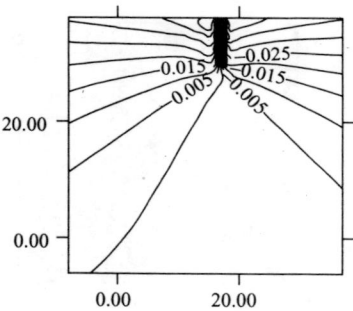

Figure 13 *Flange side, horizontal displacements, recovery data, no iterations*

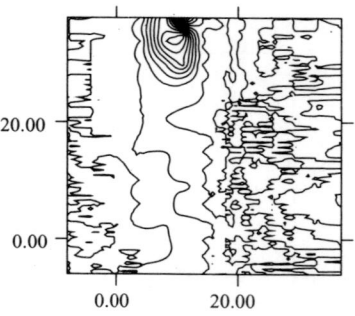

Figure 14 *Flange side, error of the horizontal displacements, no iterations, min* $= -1.47E-4$, *max* $= 1.10E-4$, *inc* $= 2E-5$

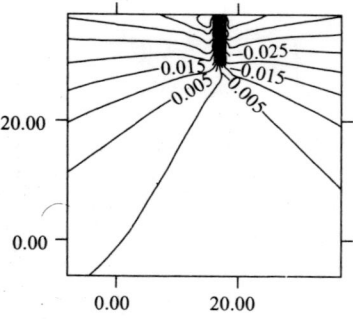

Figure 15 *Flange side, horizontal displacements, recovery data, after 7 iterations*

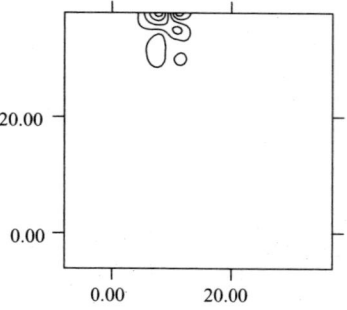

Figure 16 *Flange side, error of the horizontal displacements, after 7 iterations, min* $= -2.12E-5$, *max* $= 1.35E-5$, *inc* $= 5E-6$

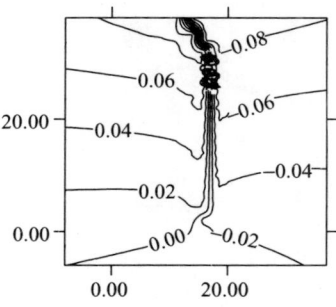

Figure 17 *Cut #5, flange side, horizontal displacements, original data, min = −9.49E−2, max = 9.00E−2*

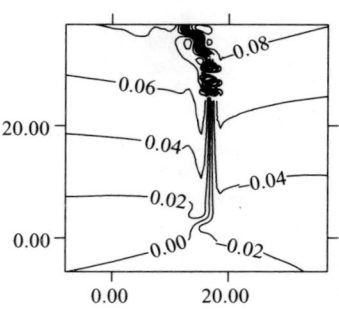

Figure 18 *Flange side, horizontal displacements, approximated data, after 7 iterations*

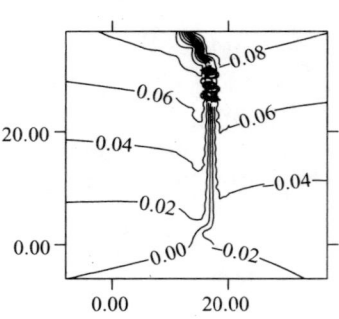

Figure 19 *Flange side, horizontal displacements, recovery data, no iterations*

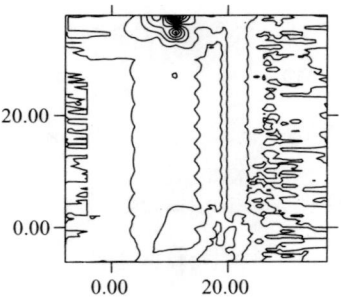

Figure 20 *Flange side, error of the horizontal displacements, no iterations, min = −3.25E−3, max = 2.72E−3, inc = 1E−3*

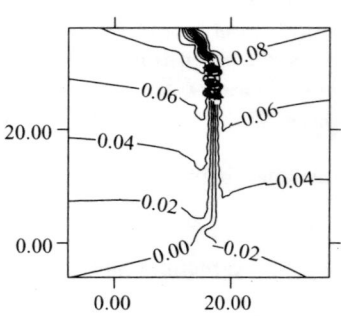

Figure 21 *Flange side, horizontal displacements, recovery data, after 7 iterations*

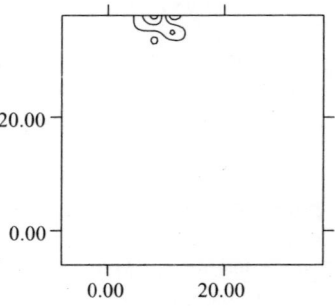

Figure 22 *Flange side, error of the horizontal displacements, after 7 iterations, min = −3.41E−4, max = 5.54E−4, inc = 2E−4*

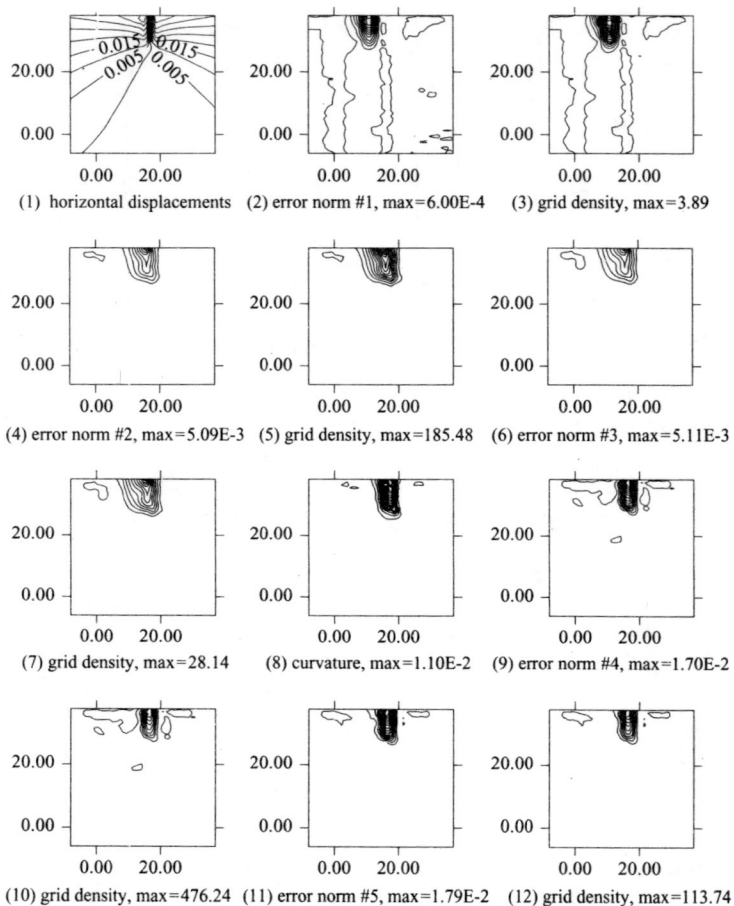

Figure 23 *Cut #3, flange side: error distribution, grid density distribution; (1) horizontal displacements; (2) error – norm #1; (3) grid density – norm #1; (4) error – norm #2; (5) grid density – norm #2; (6) error – norm #3; (7) grid density – norm #3; (8) curvature; (9) error – norm #4; (10) grid density – norm #4; (11) error – norm #5; (12) grid density – norm #5*

optim_par=optimality parameter g – mean distance between current node and central point of a star.

Results obtained are plotted in Figures 5–28. These pictures are divided into 3 different groups, namely:

Set #1 (Figures 5–22) – detailed analysis of data approximation performed for each of cuts: #1, #3, #5 of the wheel #2 for flange side (horizontal displacements) of the

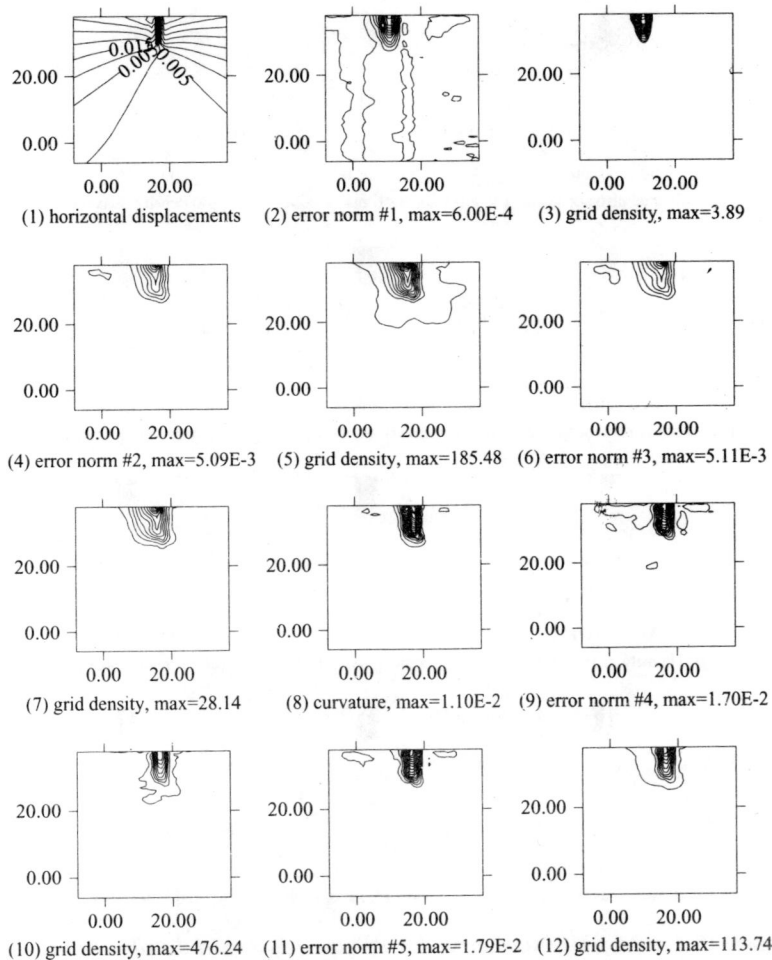

Figure 24 *Cut #3, flange side: error distribution, grid density distribution (izolines greater than 1.0 are shown); (1) horizontal displacements; (2) error – norm #1; (3) grid density – norm #1; (4) error – norm #2; (5) grid density – norm #2; (6) error – norm #3; (7) grid density – norm #3; (8) curvature; (9) error – norm #4; (10) grid density – norm #4; (11) error – norm #5; (12) grid density – norm #5*

wheel with optimal approximation parameters taken into account. Approximation parameters taken into account are: n_taylor = 8, nodes = 36, optim_par = 2.

Detailed description of the pictures is as follows:

Cut #1: Figure 5, Flange side, cut #1, horizontal displacements, original data,

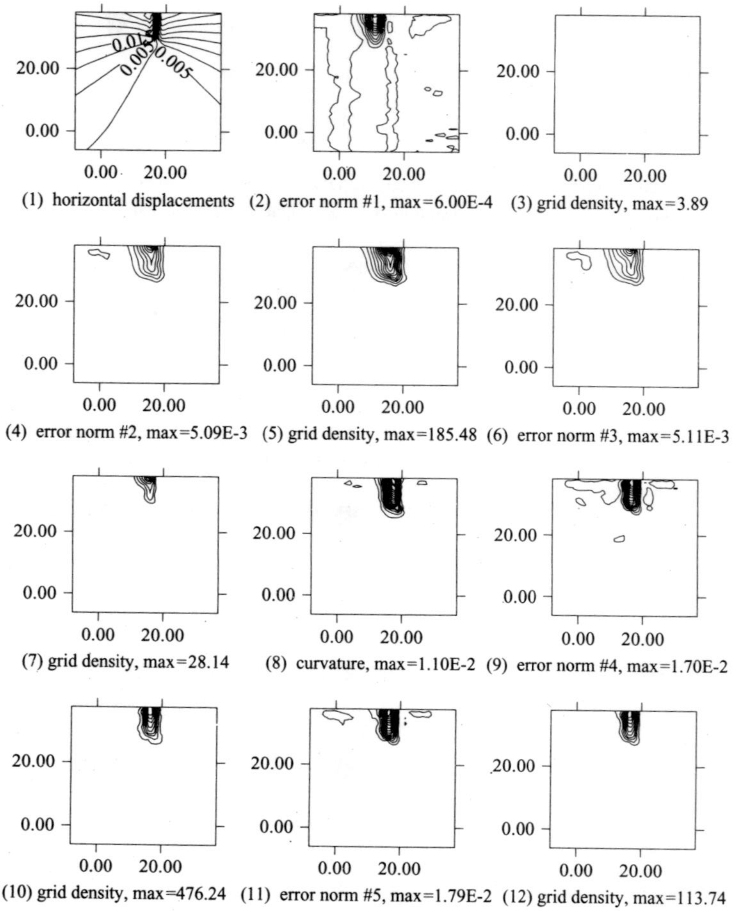

(1) horizontal displacements   (2) error norm #1, max=6.00E-4   (3) grid density, max=3.89

(4) error norm #2, max=5.09E-3   (5) grid density, max=185.48   (6) error norm #3, max=5.11E-3

(7) grid density, max=28.14   (8) curvature, max=1.10E-2   (9) error norm #4, max=1.70E-2

(10) grid density, max=476.24   (11) error norm #5, max=1.79E-2   (12) grid density, max=113.74

Figure 25 *Cut #3, flange side: error distribution, grid density distribution (izolines greater than 10.0 are shown); (1) horizontal displacements; (2) error – norm #1; (3) grid density – norm #1; (4) error – norm #2; (5) grid density – norm #2; (6) error – norm #3; (7) grid density – norm #3; (8) curvature; (9) error – norm #4; (10) grid density – norm #4; (11) error – norm #5; (12) grid density – norm #5*

Figure 6, Flange side, cut #1, horizontal displacements, approximated data after 7 iterations,

Figure 7, Flange side, cut #1, horizontal displacements, recovered data, no iterations,

Figure 8, Flange side, cut #1, error of the horizontal displacements, no iterations,

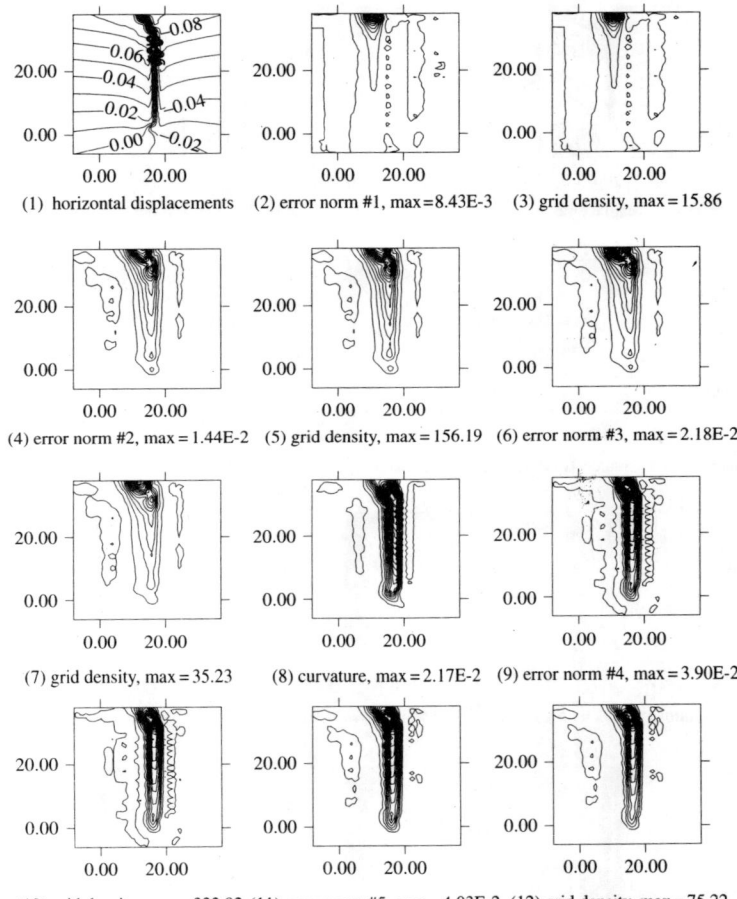

Figure 26 *Cut #5, flange side: error distribution, grid density distribution; (1) horizontal displacements; (2) error – norm #1; (3) grid density – norm #1; (4) error – norm #2; (5) grid density – norm #2; (6) error – norm #3; (7) grid density – norm #3; (8) curvature; (9) error – norm #4; (10) grid density – norm #4; (11) error – norm #5; (12) grid density – norm #5*

Figure 9, Flange side, cut #1, horizontal displacements, recovered data after 7 iterations,

Figure 10, Flange side, cut #1, error of the horizontal displacements, after 7 iterations,

Cut #3: flange side, horizontal displacements: Figures 11–16,

Cut #5: flange side, horizontal displacements: Figures 17–22.

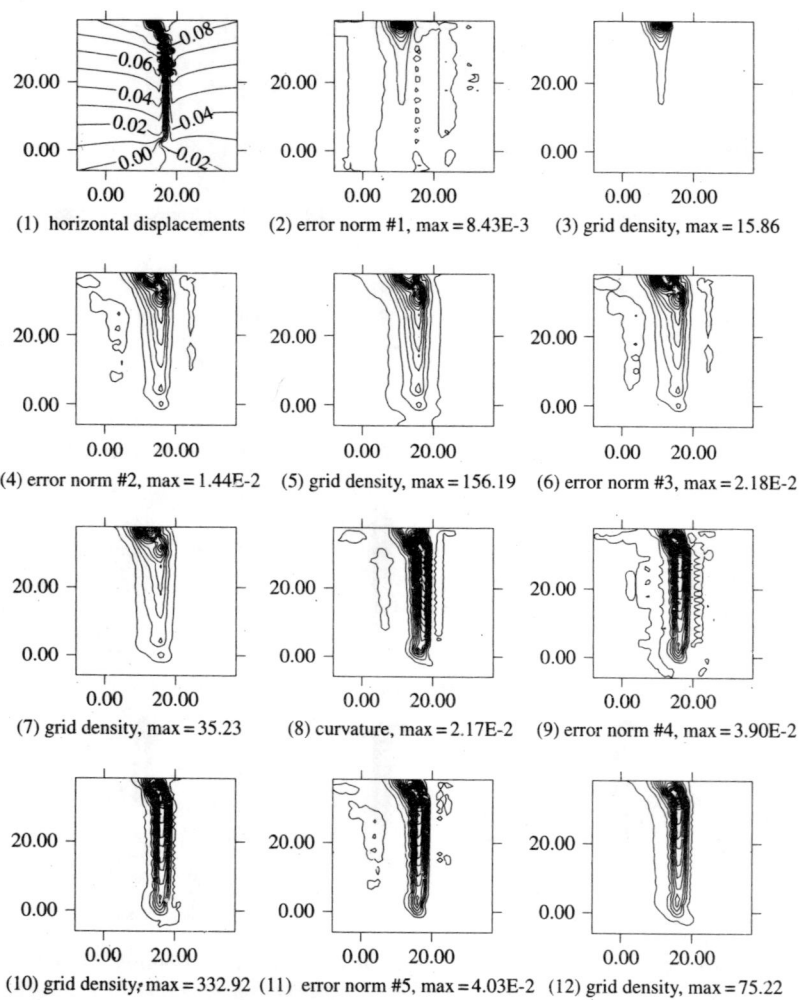

(1) horizontal displacements   (2) error norm #1, max = 8.43E-3   (3) grid density, max = 15.86

(4) error norm #2, max = 1.44E-2   (5) grid density, max = 156.19   (6) error norm #3, max = 2.18E-2

(7) grid density, max = 35.23   (8) curvature, max = 2.17E-2   (9) error norm #4, max = 3.90E-2

(10) grid density, max = 332.92   (11) error norm #5, max = 4.03E-2   (12) grid density, max = 75.22

Figure 27 *Cut #5, flange side: error distribution, grid density distribution (izolines greater than 1.0 are shown); (1) horizontal displacements; (2) error – norm #1; (3) grid density – norm #1; (4) error – norm #2; (5) grid density – norm #2; (6) error – norm #3; (7) grid density – norm #3; (8) curvature; (9) error – norm #4; (10) grid density – norm #4; (11) error – norm #5; (12) grid density – norm #5*

As one can see, because the smoothness parameter has the optimum value, approximated data is smooth enough and the errors are very small. Iterations between experimental data and fictitious data considerably decrease the errors

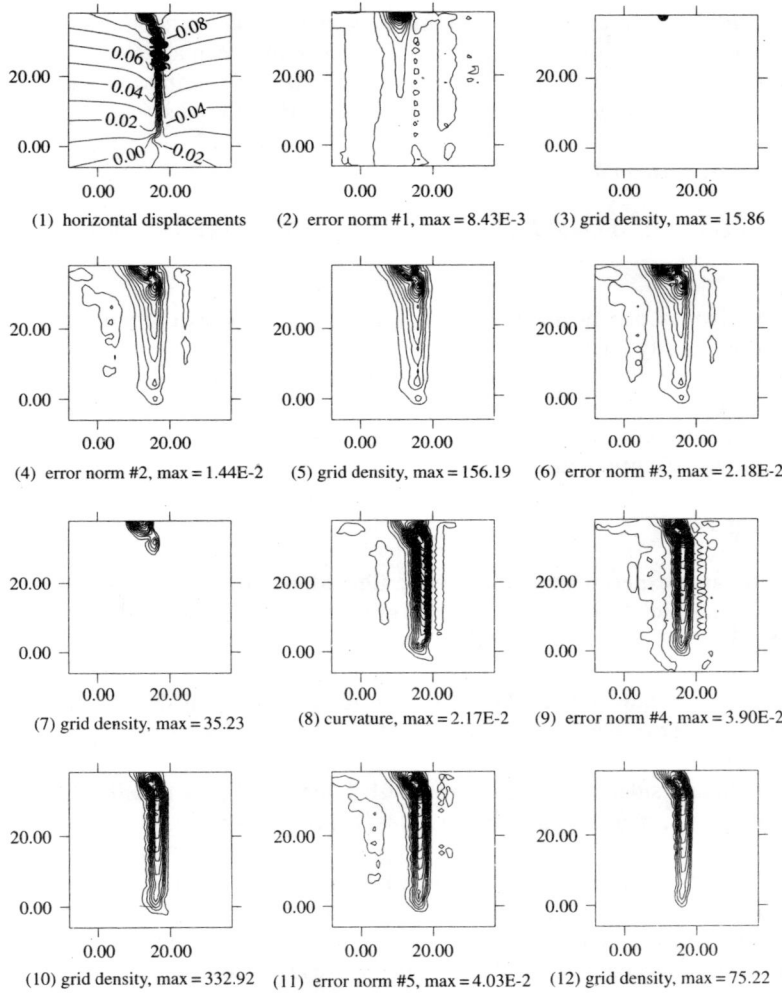

Figure 28 *Cut #5, flange side: error distribution, grid density distribution (izolines greater than 10.0 are shown); (1) horizontal displacements; (2) error – norm #1; (3) grid density – norm #1; (4) error – norm #2; (5) grid density – norm #2; (6) error – norm #3; (7) grid density – norm #3; (8) curvature; (9) error – norm #4; (10) grid density – norm #4; (11) error – norm #5; (12) grid density – norm #5*

(magnitude of the error decreases approximately 10 times). In this way one may absolutely ensure that data at experimental points and data at fictitious points are very close to each other. Thus, one may use data at fictitious points for further analysis, and this process is under error control.

Set #2, cut #3 (Figures 23–25) – summarizing pictures concerning full evaluation of the approximation process and simultaneously experimental data, flange side – horizontal displacements: local error and grid density distribution for different 'error' norms. Detailed description of the pictures is as follows:

Flange side, horizontal displacements

Cut #3, flange side, Figure 23, Error distribution and grid density distribution:

(1)   horizontal displacements, (2) error norm #1 – Sobolev norm of zero order,
(3)   grid density – norm #1,
(4)   error norm #2 – Sobolev seminorm of first order, (5) grid density – norm #2,
(6)   error norm #3 – Sobolev norm of first order, (7) grid density – norm #3,
(8)   displacement curvature, (9) error norm #4 – Sobolev seminorm of second order, (10) grid density – norm #4,
(11)   error norm #5 – Sobolev norm of second order, (12) grid density – norm #5.

Approximated experimental data may be filtered, using a certain threshold value. In Figures 24 and 25 area where required grid density is greater than 1.0% is shown for cut #3:

Cut #3, flange side, Figure 24, Error distribution and grid density distribution (izolines greater than 1.0 are shown), (1)–(12) like Figure 23,

Cut #3, flange side, Figure 25, Error distribution and grid density distribution (izolines greater than 10.0 are shown), (1)–(12) like Figure 23.

Set #3, cut #5 (Figures 26–28) – summarizing pictures concerning full evaluation of the approximation process and simultaneously experimental data, flange side – horizontal displacements: local error and grid density distribution for different 'error' norms. Detailed description of the pictures is as follows:

Cut #5, flange side, Figure 26, Error distribution and grid density distribution, – (12) like Figure 23.

In Figures 27 and 28 area where required grid density is greater than 10.0% is shown for two cuts:

Cut #5, flange side, Figure 27, Error distribution and grid density distribution (izolines greater than 1.0 are shown), (1)–(12) like Figure 23,

Cut #5, flange side, Figure 28, Error distribution and grid density distribution (izolines greater than 10.0 are shown), (1)–(12) like Figure 23.

As one may see from the pictures presented, the results strongly depend on approximation order and the smoothness parameter value. Namely, if the smoothness parameter is large – data is too smooth; this simultaneously increases the error too (but the errors are not very large, however). If magnitude of the smoothness parameter tends to smaller values, data recovered at experimental points is closer to the experimental one, but the data obtained at fictitious points is rougher.

Different error norms indicate different zones of the largest errors. Magnitudes of the norms differ essentially. Higher order norms give larger errors and are more sensitive to changes in the experimental values.

As one may observe, the zero order Sobolev norm indicates a completely different zone of the largest errors than the first or second order Sobolev norm. From Figures 23–28 one may see that cutting area is best traced by a second Sobolev semi-norm.

An error analysis has been applied to approximation of numerical data coming from FEM analysis as well. Calculation of the influence coefficient matrix (see [10]) needs approximation of numerical data from FEM mesh nodes to residual stress recovery procedure nodes (not presented here).

6. New adaptive procedure of experiment planning

As a practical result of the error analysis introduced, a new adaptive procedure of experiment planning is possible.

Experimental methods should take into account the character of the measured function; it cannot be separated from the character of measured physical field. Simply speaking, in regions where gradients of measured field are larger, one requires many more experimental points. The approach presented here provides a theoretical foundation for the above mentioned crucial condition in experimental mechanics.

One may distinguish two different situations:

1. it is possible to simulate behavior of a measured element or part of structure using a numerical method (FEM, meshless FDM),
2. it is not possible to simulate the experiment numerically.

One may note that the experiment may be or not repeated, if it sometimes one has the chance to correct the location of experimental points. If not, the approach presented defines tools for proper data evaluation and filtering.

The following procedure is proposed for the case when numerical simulation of experiment is possible:

1. Solve the problem numerically, with conditions for proper simulation of the measured part of a structure or an element as good as possible.
2. Evaluate *a posteriori* error and repeat calculation with a new mesh (grid) density, to statisfy equidistribution error requirements.
3. Define experimental grid and transfer (project) numerical solution (by means of MWLS approximation) to this grid. Try to recover original solution from experimental grid using experimental grid as a primary grid and numerical grid as a secondary grid. Evaluate *a posteriori* error and new experimental grid density function which takes into account equidistribution of an error.
4. If possible, change experimental point locations, repeat experiment and evaluate *a posteriori* error distribution (now real error).
5. Evaluate measured data using estimated error (or new required experimental grid density) as a reliability index to decide which data have to be removed or taken with lowered weight.

If the meshless method is used in above mentioned procedure, numerical simulation of the experiment is very easy, because one may directly use the experimental grid as a numerical one, without any transformations and additional (approximation) errors.

In the case when numerical simulation of the experiment is not possible, the procedure is as follows:

1. After the experiment evaluate *a posteriori* error and calculate new mesh (grid) density of experimental points with equidistribution of error.
2. If it is possible change experimental point locations, repeat experiment and evaluate *a posteriori* error distribution.
3. Evaluate measured data using error determined (or new required experimental grid density) as a reliability index to decide which data have to be removed or considered with reduced weight.

7. Final remarks

The present work is devoted to description and evaluation of the fundamental methods of the physical data approximation and the *a posteriori* error estimation *i.e.* the methods based on differences between original, experimental data (or numerical ones coming from FEM/FDM analysis) and data approximated on fictitious mesh (see [8]).

The *a posteriori* error analysis described above has been applied to the wheel saw cut data and numerical data coming from FEM analysis, using the meshless finite difference approximation. The error analysis approach presented is of great value in determination of the required concentration of experimental points in the zones where the largest stress gradients have occurred.

The current research carried out on error estimation includes:

- generalization of the Zienkiewicz–Zhu postprocessing estimator concept [17] for elastic problems in solid mechanics and its use in analysis of wheel saw cut data,
- determination of the optimal strategies for refinement of the experimental (or numerical) clouds of points, using different error norms (Sobolev norms up to second order),
- development of postprocessing techniques to enhance the solution accuracy using a different number of nodes in stars, different approximation order (*i.e.* 2nd or 3rd order) and an additional iterative process to smoothen the largest discrepancies between data on original (experimental) and fictitious (numerical) grids,
- formulation of the new *adaptive approach to experiment planning and implementation*, taking into account *a posteriori* error estimation and distribution of experimental points with equidistributed error,
- analysis of wheel saw cut data, especially for wheel #2 (see R. Czarnek [3]), 5 cuts of the wheel, both flange and 2nd sides of the wheel analyzed,
- analysis of numerical data, coming from FEM analysis, for wheel #2 [11].

Advantages of the error analysis performed on the experimental as well as numerical data (see FEM/FDM analysis [11]) have been shown. A significant step towards a new adaptive analysis (approximation) of the physical data was made. Besides, the approach presented here yields formulation of new requirements against measurement devices possible, thus making way for adaptive experimental data collection.

The proposed further research includes: development of reliable error estimates for computed "physical fields" with the efficiency index close to 1 (approximated fields are very close to original ones), further development of the optimal strategies for 'h' adaptive refinement of the experimental data points cloud, development of adaptive modeling in which certain features of physical models are incorporated and stress analysis of deformation fields in rails and wheels.

8. References

1. Belytschko T., Krongaus Y., Organ D., Flemming M., Krysl P., "Meshless Methods: An Overview and Recent Development", *Comp. Meth. in Appl. Mech. and Engng*, 139, 3–44, 1996.

2. Bugeda G., Onate E., "New Adaptive Techniques for Structural Problems", *Numerical Methods in Engng'92*, Ch.Chirsh *et al.* (Editors), Elsevier Science Publishers B.V., 1992.

3. Czarnek R., *Experimental Determination of Release Fields in Cut Railroad Car Wheels*, DOT/FRA/ORD-96/DOT-VNTSC-FRA-96, Final Report, Cambridge, USA, October, 1996.

4. Demkowicz L., Oden J.T., Rachowicz W., Westerman T.A., "Toward a Universal h-p Adaptive Finite Element Strategy. Part2: A Posteriori Error Estimation", *Computer Methods in Applied Mechanics and Engineering*, 77(1–2), 113–180, 1989.

5. Karmowski W., Orkisz J., "Physicall Based Enhanced Analysis of Stresses Using Experimental Data", in: *Quality and Maintenance for Modern Railway Operation*, editors J.J. Kalker *et al.*, 287–296, Delft, June 24–26, 1992.

6. Karmowski W., Orkisz J., "Physically Based Method of Enhanced of Experimental Data – Concept, Formulation and Application to Identification of Residual Stresses", *Proc. of the IUTAM Symposium on Inverse Problems in Engng Mechanics*, May 11–15 Tokyo, Japan, Springer-Verlag, 61–70, 1993.

7. Krok J., Orkisz J., "Application of the Generalized FDM to Calculation of Arbitrary Loaded Axisymmetrical Massive Structures", *Proc of 28-th Conf. KILiW PAN and KN PZITB*, Krynica, Poland, 1982, 81–90 (in polish).

8. Krok J., New Approach of Error Control in Approximation and Smoothing of Physical Data, Application to Wheel Saw Cut Measurements Data, Report to the VNTSC, Cambridge, USA, 1998.

9. Krok J., Orkisz J., "Application of the Generalized FD Approach to Stress Evaluation in the FE Solution", *Int. Conf. on Comp. Mech.*, Tokyo 1986, XII, 31–36.

10. Krok J., Orkisz J., "A Unified Approach to the FE Generalized Variational FD Method in Nonlinear Mechanics, Concept and Numerical Approach", in: *Discretization Method in Structural Mechanics, IUTAM/IACM Symposium Vienna 1989*, 353–362, Springer-Verlag, 1990.

11. Krok J., Orkisz J., Skrzat A., "Reconstruction of Hoop Stresses in 3D Bodies of Revolution Based on Simulated Saw Cut Data", *XIII Conf. on Comp. Meth. in Mechanics*, Poznań, Poland, 669–676, 1997.

12. Krok J., Orkisz J., "Unified Approach to the Adaptive FEM and Meshless FDM. Concept and Tests", 2nd *European Conference on Computational Mechanics*, June 26–29, Cracow, Poland , 2001, 1–33.

13. Liszka T., Orkisz J., *The finite difference method at arbitrary irregular grids and its applications in applied mechanics*, Computers and Structers, 11, 83–95, 1980.

14. Liszka T., "An interpolation method for an irregular net of nodes", *International Journal for Numerical Methods in Engineering*, 20, 1599–1612, 1984.

15. Orkisz J., "The Finite Difference Method", Part III, in: *Numerical Methods in Mechanics*, in: Springer-Verlag, 1998.

16. Zhu J.Z., Hinton E., Zienkiewicz O.C., "Mesh Enrichment Against Mesh Regeneration Using Quadrilateral Elements", *Comm. in Num. Meth. in Engng*, Vol. 9, 547–554, 1993.

17. Zienkiewicz O.C., Zhu J.Z., "A Simple Error Estimator and Adaptive Procedure for Practical Engineering Analysis", *Int. Journ. Num. Meth. Eng.*, 24, 337–357, 1987.

18. Zienkiewicz O.C., Zhu J.Z., "The Superconvergent Patch Recovery and *a posteriori* Error Estimates", Part 2: Error Estimates and Adaptivity, *Int. J. Num. Meth. Engng*, Vol. 33, 1365–1382, 1992.

19. Zienkiewicz O.C., Taylor R.L., *Finite Element Method*, Butterworth, Oxford, 2000.

20. Gavette L., Cuesta J.L., Ruiz A., "A Procedure for Approximation of the Error in the EFG Method", *Int. Journ. Num. Meth. Eng.*, Vol. 53, 677–690, 2002.

21. Skrzat A., Orkisz J., Krok J., "Residual Stress Reconstruction in Railroad Car Wheels Based on Experimental Data Measured at Saw Cut Test", 2nd *European Conference on Computational Mechanics*, June 26–29, Cracow, Poland , 2001, 1–17.

Chapter 5

A Meshless Approach for 2D Vibro-Acoustic Problems

Philippe Bouillard, Valéry Lacroix & Eric De Bel
Université Libre de Bruxelles
Department of Continuum Mechanics, Belgium

1. Introduction

In recent years, there has been increasing interest in simulation of noise, either to satisfy code rules or to improve the end-user's comfort. Here, we consider the noise generated by structural vibrations. The numerical methods usually used are, for low frequencies, the *finite element method* (FEM), coupled if necessary with a numerical treatment for *infinite domains* (IFEM or DtN), or the boundary element method (BEM). For high frequencies, these deterministic approaches are no more suited and *statistical energy analysis* is the most popular solution. Between the low and high frequency regions, *i.e.* for so-called medium frequencies, all the methods have been attempted but none of them seems accurate enough for engineering purposes.

The frequency range within which the standard finite element method can be applied is limited to the *a priori* error estimates results. It is well known today that the resolution rule (rule of the thumb) discretising a wavelength by a given number of elements (typically 6 to 10) is not sufficient because it does not take the pollution effect into account. The error estimate for the linear finite element method has been proved by F. Ihlenburg *et al.* for the uncoupled acoustic problem [IHL 95]

$$\|p - p_h\| \le C_1 kh + C_2 k^3 h^2 \tag{1}$$

In practice, this result restricts the finite element analysis to a few hundred Hertz for the acoustic simulation of a car cabin. It is of course not sufficient because the human ear is sensitive mostly up to 2000 Hz [BOU 98a]. There is then a need for a reliable numerical method allowing engineering computations in the frequency range [0–2000] Hz for any geometry.

This problem is one the most challenging problems that the researchers try to address today. Lots of solutions have been proposed, firstly based on the idea of stabilizing the finite element method itself.

High order approximations have also been proposed, such as the *hp*-FEM by L. Demkowicz [GER 96], the *reproducing kernel particle method* (RKPM) by W. K. Liu [URA 97, VOT 01] or the element-free Galerkin method proposed by T. Belytschko [BEL 94, BEL 96] extended by Ph. Bouillard to acoustics [BOU 98b].

These methods are already interesting but everybody seems to agree that it is even more advantageous to use a set of plane wave solutions, like E. Chadwick and P. Bettes [CHA 97] or Ch. Farhat, I. Harari and L. P. Franca who formulate a discontinuous Galerkin FEM [FAR 00], or even to built the subspace by including terms of the solution of the homogeneous equation. A natural and very efficient way to achieve this is to use generalized formulations, like I. Babuška and J. M. Melenk [BAB 97] with their very popular partition of unity method (PUM) or to use a Trefftz subspace [DES 98].

Most of these approaches are approximation methods (not interpolation) and result in highly oscillating shape functions. Two major drawbacks have to be overcome: essential boundary conditions and numerical integration. Most of them have also been developed for uncoupled acoustics. However, improving simulation of structural vibration seems the highest priority.

The paper aims to show how a meshless approach really constitutes a competitive deterministic approach up to medium frequencies for coupled vibroacoustic problems. Based on our previous work, we suggest using an improved EFGM for the fluid domain and a generalized FEM for the structural domain.

The paper is organized as follows: section 2 addresses the formulation of the vibroacoustic problem, section 3 is dedicated to the discretisation of the fluid domain by an improved element-free Galerkin method and section 4 provides a partition of unity approach to solve the structural dynamic problem. Finally, conclusions are given in section 5.

2. Vibroacoustic model problem

Consider an elastic solid domain Ω_s coupled with a fluid domain Ω_f along a wet surface Γ (Figure 1). Within the solid, we assume that the displacements u_i are small perturbations around a steady state and, at a first approach, we neglect the structural damping and the body forces. In the fluid domain, we assume that the acoustical wave propagates harmonically in a non viscous without body forces fluid around a steady state (linear acoustics). We do not consider the acoustic damping

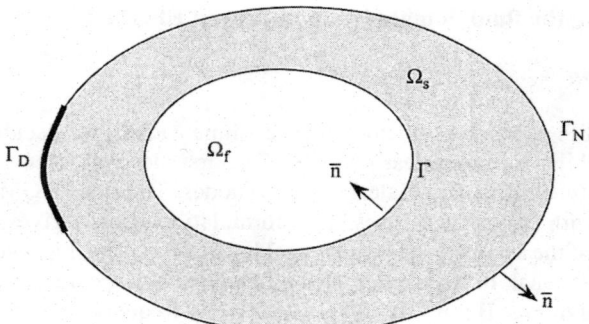

Figure 1 *A coupled vibro-acoustic domain*

yet. Then, the problem is addressed by the system of equations (2), where ρ_s and ρ_f denote the mass density of the solid and of the fluid respectively.

$$
\begin{aligned}
\sigma_{ij,j}(u) + \rho_s\omega^2 u_i &= 0 && \text{in } \Omega_s && (a)\\
\sigma_{ij}(u)n_j &= \bar{F}_i && \text{on } \Gamma_N && (b)\\
u_i &= \bar{u}_i && \text{on } \Gamma_D && (c)\\
\sigma_{ij}(u)n_j &= -pn_i && \text{on } \Gamma && (d)\\
\tfrac{\partial p}{\partial n} &= \rho_f\omega^2 u_i n_i && \text{on } \Gamma && (e)\\
\Delta p + \tfrac{\omega^2}{c^2}p &= 0 && \text{on } \Omega_f && (f)
\end{aligned}
\tag{2}
$$

Equation 2(a) is the elastodynamic classical equation, with its boundary conditions 2(c) on Γ_D (restraints) and 2(b) on Γ_N (tractions). Equation 2(d) represents the action of the pressure forces on the structure. Equation 2(e) represents the action of the structural vibrations on the fluid. If the structural velocities are given, the problem is said to be uncoupled or weakly coupled, and equation 2(e) is reduced to a Neumann boundary condition for the fluid. Finally, equation 2(f) is the Helmholtz equation for the acoustic pressure p. Further information about fluid-structure interactions can be found in [MOR 95].

Whatever the approximation method, the discretization of the variational form leads to the linear system of equations

$$
\begin{bmatrix} K_s - \omega^2 M_s & K_{sf} \\ K_{fs} & K_f - \omega^2 M_f \end{bmatrix}
\begin{Bmatrix} u \\ p \end{Bmatrix} = \begin{Bmatrix} f \\ 0 \end{Bmatrix}
\tag{3}
$$

where K_s and M_s are the structural stiffness and the structural mass matrix respectively, K_f and M_f are the acoustic stiffness and the acoustic mass matrix, K_{sf} and K_{fs} the coupling matrices. This formulation is not symmetrical.

3. Discretizing the fluid domain by an improved EFGM

3.1. *Motivation*

As most real-life acoustic problems are three-dimensional, we decided to investigate the possibilities of meshless methods. Our first idea was to couple structural
finite element to element-free Galerkin (EFG) nodes. The first EFGM that we formulated was with polynomial bases. This formulation already gives a significant
improvement of the accuracy vs. the classical linear FEM. The improvement comes
from the non rational, or high order, shape functions better suited to approximate
waves than polynoms. But it was still limited to low frequencies, so we decided to
look for a basis that takes the wave propagation phenomenon into account, first by
putting in the basis a set of plane waves, then by accepting the idea that the basis can
be locally defined. Numerically, this is achieved by an iterative defect-correction
type method.

For the particular case of the Helmholtz equation $2(f)$, we take advantage of the
fact that the local basis of an element-free Galerkin method can naturally contain
terms which are a solution of the Helmholtz equation. In acoustics, as the pressure
is a complex variable, terms in $\cos \theta(x, y, z)$ and $\sin \theta(x, y, z)$ are introduced in
the *meshless* basis, where $\theta(x, y, z)$ is the value of the phase of the pressure field
at each point (x, y, z) of the domain. Since $\theta(x, y, z)$ is *a priori* unknown, it has
first to be computed for instance with a polynomial linear basis. Then, with the
new θ-dependant local *meshless* basis, very accurate results are demonstrated on
academic and real-life 3D problems within a large frequency range.

3.2. *Formulation*

The pressure, which is a complex variable, can always be written as

$$p(x, y) = \bar{P}(x, y)[\cos \theta(x, y) + \mathrm{j} \sin \theta(x, y)] \tag{4}$$

where $\bar{P}(x, y)$ is the amplitude of the wave and $\theta(x, y)$ its phase. Therefore, if
the phase is exactly known over the whole domain, then the approximate pressure p^h (the upper h standing for numerical solution) can be *exactly* computed by
considering an expansion

$$p^h(x) = P^t(x) a(x) \tag{5}$$

with the basis

$$\mathbf{P}^t(x, y) = \{1, \cos \theta(x, y), \sin \theta(x, y)\} \tag{6}$$

where unknown coefficient $a(x)$ are fixed by using a moving least square
approximation.

Obviously, for real-life cases, the distribution of $\theta(x, y)$ is *a priori* unknown. Thus, in the latter, $\theta(x, y)$ will be approximated by a distribution $\theta^h(x, y)$ obtained by a first computation of the pressure field using, for instance, a linear polynomial *meshless* basis

$$\boldsymbol{P}^t(x, y) = \{1, x, y\} \tag{7}$$

With this basis, a first approximation of the pressure and of the phase is computed

$$\cos \theta_I^h = \frac{p_{rI}^h}{\sqrt{\left(p_{rI}^h\right)^2 + \left(p_{iI}^h\right)^2}} \quad \text{and} \quad \sin \theta_I^h = \frac{p_{iI}^h}{\sqrt{\left(p_{rI}^h\right)^2 + \left(p_{iI}^h\right)^2}} \tag{8}$$

where the pressure is split into its real and imaginary part.

$$p_I^h(x, y) = p_{rI}^h(x, y) + j p_{iI}^h(x, y) \tag{9}$$

Then, consider the basis defined by

$$\boldsymbol{P}^t(x, y) = \{1, \cos \theta_I^h(x, y), \sin \theta_I^h(x, y)\} \tag{10}$$

with $\cos \theta_I^h(x, y)$ and $\sin \theta_I^h(x, y)$ coming from the first computation and compute a new approximated pressure field $p_{II}^h(x, y)$. Of course, this method can be iterated: a third approximation of the pressure can be computed by building a basis of type (6) with equations (8) but by using $p_{II}^h(x, y)$ instead of $p_I^h(x, y)$ and so on until the correction on θ will satisfy a tolerance criterion.

3.3. Numerical examples

Consider the academic example of a plane wave propagating in a cubic cavity (Figure 2).

The analytical solution of this problem is known and given by:

$$\begin{aligned} p(x, y, z) = {} & \cos k(x \cos \alpha \cos \beta + y \sin \alpha \sin \beta + z \cos \beta) \\ & + j \sin k(x \cos \alpha \cos \beta + y \sin \alpha \sin \beta + z \cos \beta) \end{aligned} \tag{11}$$

where α and β define the propagation direction.

The FRF is computed using linear FEM, linear basis EFGM and the defect-correction method limited to one iteration. The analytical FRF is also represented. These curves are shown in Figure 3 for the real part of the pressure. The lower and upper bounds of the frequencies are 1 Hz and 1500 Hz. The response is given in dBA.

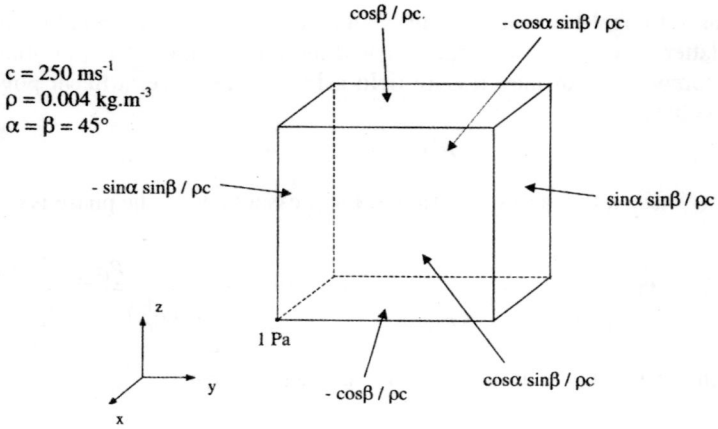

Figure 2 *Cubic cavity-plane wave propagation*

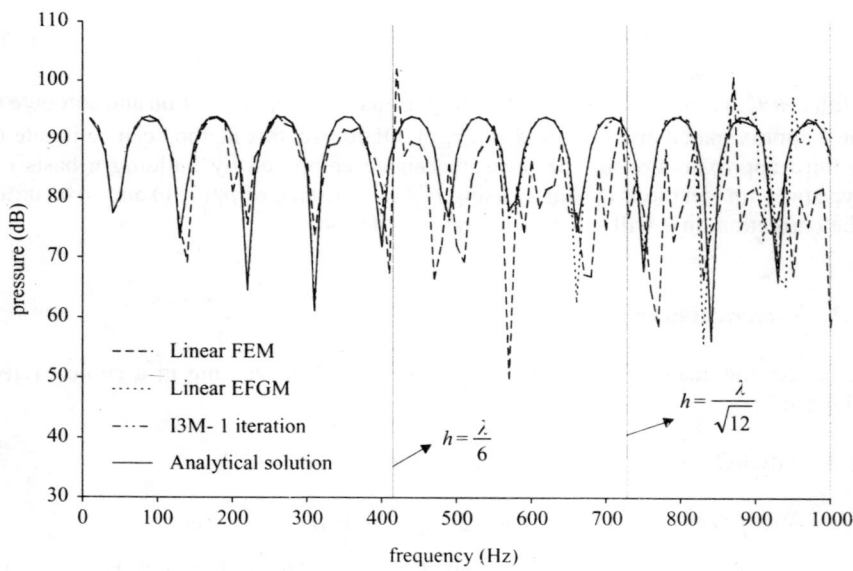

Figure 3 *FRF for the real part of the pressure in the middle of the cube*

One can notice that the defect-correction method presents very good behaviour when the frequency increases over the numerical description limit of the wave with linear FEM [BOU 98a] *i.e.* $h = \lambda/\sqrt{12}$. The frequency corresponding to the classical rule of the thumb for linear FEM has also been plotted *i.e.* $h = \lambda/6$.

4. Discretizing the structural domain by a GFEM

4.1. *Motivation*

The motivation to develop a meshless solution for the elastodynamic problem is based on the same reasons as for the acoustic problem. Here again, a formulation able to capture the wave propagation phenomena, particularly for medium frequencies, is needed. However, since most of the real-life vibroacoustic problems are three dimensional, it is necessary to formulate a method for the shell behaviour. Even if there are some trials to formulate EFGM for the shell problem [KAN 01], it seems to us more efficient to keep the geometric information of the shells with (finite) elements and, at the same time, to improve their formulation by taking advantage of the meshless solution. A very popular and easy way to achieve this is the partition of unity method, proposed by I. Babuška [MEL 97] and extended by T. Belytschko [MOE 02].

The partition of unity method can be seen as a generalized finite element method where the core ideas are, first, the construction of the spaces with local approximation properties and, second, the conformity of these spaces. Then, a feature of those spaces is that it can approximate well the exact solution locally. When the exact solution of a problem can be expressed, the PUM can give very accurate results [MEL 97]. If it is not the case, the introduction of other functions in the space looks like a p-refinement of the finite element solution. In this paper, the PUM is formulated with a local enrichment of the basis based on the exact solution of the elastodynamic problem. The terms of the exact solution of the homogeneous problem are put in the local basis everywhere since the pollution of wave propagation problem is global.

The foundations of the PUM consist in partitioning the unity. Consider a set of functions N_i and a domain Ω overlapped by a set of open domains Ω_i, the so-called *patches*, such as:

$$\text{supp}(N_i) = \Omega_i$$
$$\forall x \in \Omega, \sum_i N_i = 1 \tag{12}$$

where $\text{supp}(N_i)$ denotes the support of definition of the function N_i.

The N_i comprise the partition of unity attached to the patch Ω_i. Consider now the space of functions V_i^p defined on Ω_i. In this case, the space of functions used for the approximation is

$$V = \text{span}\left\{N_i v_i^p\right\} \qquad \text{with } v_i^p \in V_i^p \tag{13}$$

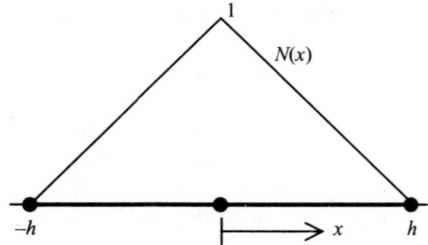

Figure 4 *Hat function*

where span $\left\{N\, v_i^p\right\}$ denotes the space of functions generated by the set of functions $N_i v_i^p$.

Each node has several degrees of freedom (one per function of V_i^p) and the approximation of a function at the point x is given by:

$$u^h(x) = \sum_i \sum_{v_i^p \in V_i^p} a_{i,p} N_i v_i^p(x) \qquad (14)$$

The $a_{i,p}$ are the unknown coefficients and they can be computed either using a collocation method or a Galerkin method. To formulate the PUM as an enriched FEM, one has to correctly choose the patches Ω_i. For instance, for beam problems, the patch contains two adjacent finite elements (the patches Ω_i are overlapping) and the functions N_i can be the usual hat functions.

The main advantages of the PUM are:

– the introduction of *a priori* known terms in the base V is possible;
– the shape functions are easily computed in comparison with other p-methods;
– spaces of any desired regularity can be constructed. Then, the *test* functions needed in the variational formulations of high order differential equations (such as beam and shell problems) become available.

However, major drawbacks have to be overcome:

– the numerical integration of high order functions require appropriate schemes;
– the continuity of the displacement field of non coplanar shells requires specific attention. In a PUM, it is even more necessary since the unknowns are not the displacement components but the coefficients of the subspace expansion.

4.2. *Formulation*

4.2.1. The improved Timoshenko element

As an example, consider the case of the Timoshenko beam. Then, the general solution of the second order differential equation for the longitudinal displacement is given by :

$$u(x) = A \cos(kx) + B \sin(kx) \quad \text{with } k = \sqrt{\frac{\rho \omega^2}{E}} \tag{15}$$

and the general solution of the fourth order differential equation for the transverse deflection is given by:

$$w(x) = C_1 \sin\left(\lambda_1 \frac{x}{L}\right) + C_2 \cos\left(\lambda_1 \frac{x}{L}\right) + C_3 sh\left(\lambda_2 \frac{x}{L}\right) + C_4.ch\left(\lambda_2 \frac{x}{L}\right) \tag{16}$$

where λ_i are non dimensional parameters function of the material and the geometrical properties of the beam and of the pulse.

Figure 5 shows the degrees of freedom of the element where u_i are the longitudinal displacements, w_i the transverse deflections and θ_i the rotations.

The approximated corresponding fields are

$$u^h(x) = \sum_i N_i^u(x) \sum_j V_j^{u(i)}(x) \cdot a_{ij}^u = \sum_i \{\Phi_i^u\}^t \{A_i^u\}$$

$$w^h(x) = \sum_i N_i^w(x) \sum_j V_j^{w(i)}(x) cdot a_{ij}^w = \sum_i \{\Phi_i^w\}^t \{A_i^w\} \tag{17}$$

$$\theta^h(x) = \sum_i N_i^\theta(x) \sum_j V_j^{\theta(i)}(x) \cdot a_{ij}^\theta = \sum_i \{\Phi_i^\theta\}^t \{A_i^\theta\}$$

where $\{\Phi_i\}^t = \begin{bmatrix} N_i V_1^{(i)} & \dots & N_i V_m^{(i)} \end{bmatrix}$ and $\{A_i\}^t = \{a_{i1} \quad \dots \quad a_{im}\}$ (m is the number of functions in the basis). The basis for the approximations of w or θ are chosen different in order to avoid the shear locking problem, as in [korn]. According to equations (15–16), the basis contains {sin,cos} or {sin,cos,sh,ch} functions.

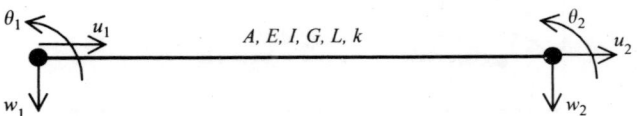

Figure 5 *Timoshenko beam element*

4.2.2. Trusses of beams

The formulation of the Timoshenko element for co-linear beams is really natural. It is no more the case when considering trusses of non co-linear beams. The formulation in a general global system of axes, which is easily formulated for finite elements, seems to present additional difficulties when the unknowns are not the displacement field components but the coefficients of a non polynomial expansion different for the u and w components. In Figure 6, the global axes are represented by capital letters whereas the local axes are always represented by lower-case letters.

From (17), the unknowns in the local system of axes are:

$$\begin{cases} u = f(\sin(kx), \cos(kx)) \\ w = f(\sin(\lambda_1 x), \cos(\lambda_1 x), \sinh(\lambda_2 x), \cosh(\lambda_2 x)) \end{cases} \tag{18}$$

In the global axes, they become:

$$\begin{cases} U = f(u, w) = f(\sin(kx), \cos(kx), \sin(\lambda_1 x), \cos(\lambda_1 x), \sinh(\lambda_2 x), \cosh(\lambda_2 x)) \\ W = f(u, w) = f(\sin(kx), \cos(kx), \sin(\lambda_1 x), \cos(\lambda_1 x), \sinh(\lambda_2 x), \cosh(\lambda_2 x)) \end{cases} \tag{19}$$

The 10 d.o.f.s per node system (2 for u, 4 for w and 4 for θ) in local axes, become a 16 d.o.f.s per node system (6 for u and w and still 4 for θ). The problem becomes still more complex when two beams with different orientation are connected. Indeed, the global displacements of the node are a combination of the local displacements of each beam connected to this node. An easy but costly way to implement the assembly of non co-linear beams is to introduce a set of linear constraints between the d.o.f.s (Figure 7) by Lagrange multipliers.

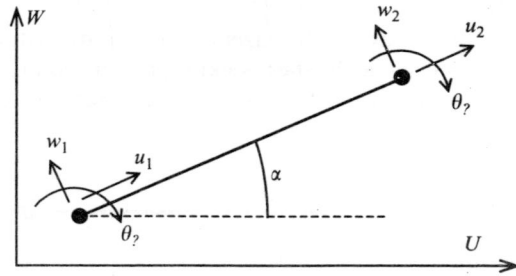

Figure 6 *Timoshenko beam element in the global axes*

Figure 7 *Constraints to connect duplicated nodes*

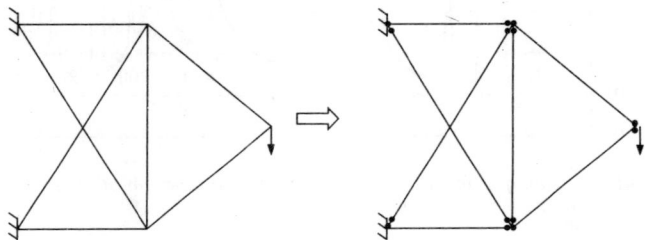

Figure 8 *Truss of Timoshenko beams under dynamic loading*

4.3. *Numerical tests*

Within this paper, we will focus on the two major aspects that must be resolved with the proposed formulation: numerical quadrature and continuity for non co-linear beams. We consider a simple truss with rigid nodes, loaded by a vertical dynamic force (Figure 8). Each beam is discretised by only one element. Boundary conditions and constraints are introduced by the Lagrange multiplier technique.

4.3.1. Comparison FEM-PUM

As a first test, a comparison is performed between high order finite elements ($p = 3$, 183 d.o.f.s) with the solution obtained with the PUM formulation incorporating analytical terms into the basis (173 d.o.f.s). Figure 9 shows that the PUM solution is more accurate: for increasing frequencies, the PUM solution better captures the oscillating waves.

4.3.2. Numerical quadrature

First, we will study the influence of the numerical quadrature on the accuracy of the solution when considering the basis containing analytical terms (PUM-ex). We consider a reference numerical solution with 100 Gauss points to overkill the numerical error. Then, we compare the solution obtained with 4, 5 and 6 Gauss points. Figure 10 shows that the error on the numerical scheme is too large for

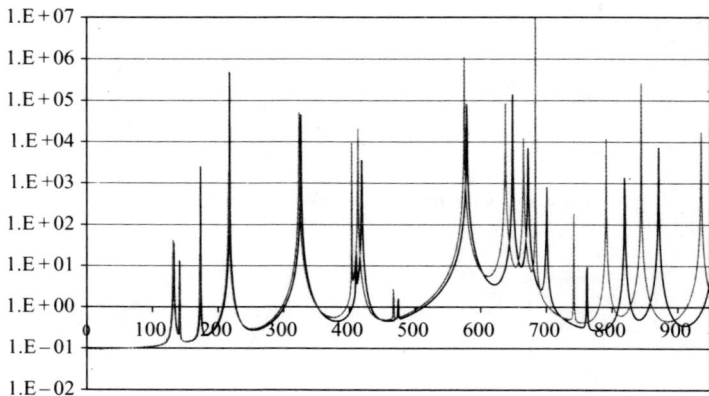

Figure 9 *Comparison FEM-PUMex: error as a function of the frequency*

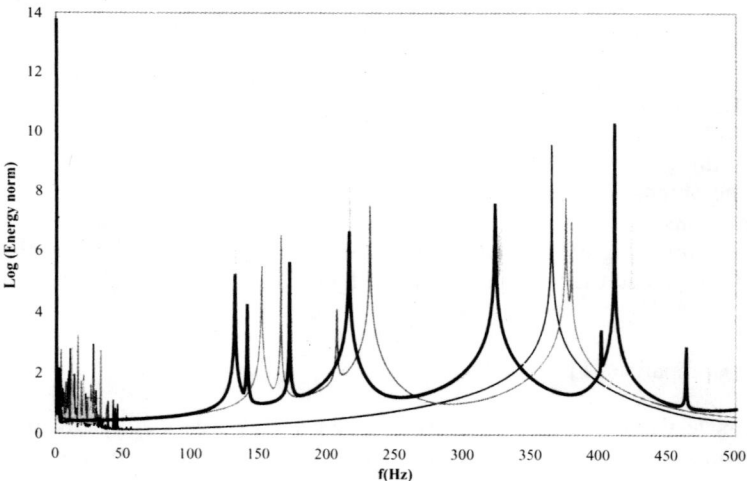

Figure 10 *Influence of the number of Gauss points*

a small number of Gauss points. It seems necessary to look for a more suitable integration method.

Secondly, we study the influence on the eigenfrequencies computations and also compare our results with those obtained with a polynomial basis (PUM-poly/n, where n denotes the number of elements discretizing a beam, the basis is always composed of monomials up to order 3).

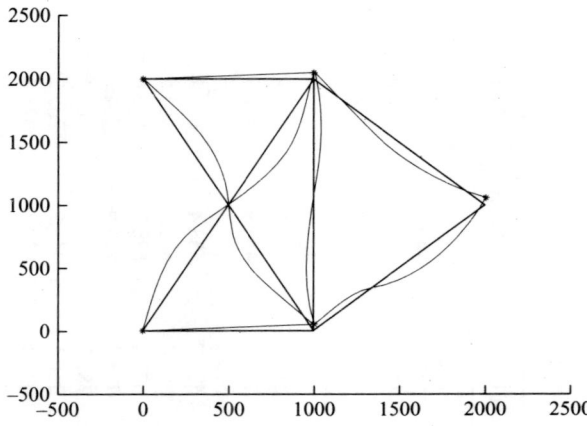

Figure 11 *Seventh eigenmode*

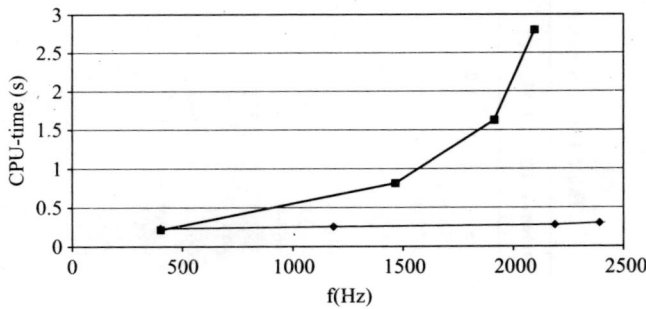

Figure 12 *PUMpoly vs PUMex: cutoff frequency as a function of the frequency*

As expected, Table 1 shows that the accuracy of the solution increases with the quality of the integration. For medium frequencies, it will be necessary to improve the numerical integration scheme. As an example of eigenmodes, the seventh mode is plotted in Figure 11.

However, there exists a main advantage in using an analytical base. When discretizing with classical *hp*-versions of the FEM, called here PUMpoly, medium frequencies require refined mesh. It is not the case with PUMex where the number of d.o.f.s is fixed, even if a fine integration scheme is needed. Figure 12 shows the cutoff frequency (the frequency above which the wave is numerically damped) as a function of the computational time. It shows that the PUMex method exhibits an almost constant time.

Table 1 *Eigenfrequencies*

No	Référence	PUMex-6	PUMex-8	PUMex-10	PUMex-12	PUMpoly1	PUMpoly2	PUMpoly3	PUMpoly4
1	131.6	132.2	131.6	131.6	131.6	131.9	131.6	131.6	131.6
2	140.7	141.4	140.7	140.7	140.7	141.0	140.7	140.7	140.7
3	172.0	173.3	172.0	172.0	172.0	173.3	172.0	172.0	172.0
4	215.7	216.0	215.7	215.7	215.7	216.0	215.7	215.7	215.7
5	322.9	325.7	322.9	322.9	322.9	327.3	323.2	322.9	322.9
6	401.3	420.6	401.3	401.3	401.3	543.3	401.3	401.3	401.3
7	410.8	480.0	410.8	410.8	410.8	567.6	411.1	410.8	410.8
8	463.9	579.0	463.9	463.9	463.9	633.4	463.9	463.9	463.9
9	573.3	646.7	573.3	573.3	573.3	673.6	575.2	574.3	573.6
10	633.1	665.3	633.1	633.1	633.1	766.9	633.7	633.4	633.1
11	662.2	703.6	662.2	662.2	662.2	889.2	663.4	662.8	662.5
12	676.4	857.3	676.4	676.4	676.4	1040.1	677.0	676.7	676.4
13	734.6	947.1	734.9	734.6	734.6		735.2	734.9	734.6
14	781.1	1055.9	781.1	781.1	781.1		781.7	781.1	781.1
15	834.5	1370.2	834.5	834.5	834.5		836.1	835.2	835.2
16	927.8		927.8	927.8	927.8		934.1	930.3	928.8
17	1004.0		1004.3	1004.0	1004.0		1008.5	1005.6	1004.7
18	1035.0		1034.7	1035.0	1035.0		1042.6	1037.9	1036.0
19	1185.5		1048.0	1185.5	1185.5		1189.3	1186.8	1186.5
20	1230.1		1165.6	1229.8	1230.1		1232.3	1230.8	1230.4

21	1353.5	1341.1	1353.1	1353.5	1362.6	1356.3	1355.0
22	1465.4	1455.9	1465.1	1465.4	1495.	1478.4	1472.7
23	1578.3		1578.0	1578.3		1581.5	1579.9
24	1584.0		1584.0	1584.0		1591.9	1588.4
25	1651.7		1651.7	1651.7		1654.2	1652.9
26	1683.6		1683.9	1683.6		1695.3	1690.6
27	1691.2		1691.2	1691.2		1695.9	1693.4
28	1757.9		1760.4	1758.2		1762.0	1760.1
29	1841.7		1843.3	1842.0		1848.0	1844.9
30	1914.4		1911.6	1914.8		1935.0	1925.8
31	2096.6		2082.4	2096.6		2126.6	2113.0
32	2188.6		2116.2	2189.6		2197.5	
33	2274.6		2135.2	2274.6		2311.9	
34	2369.8		2194.3	2372.7			
35	2391.9			2589.3			
36	2593.4			2680.7			
37	2681.0			2821.7			

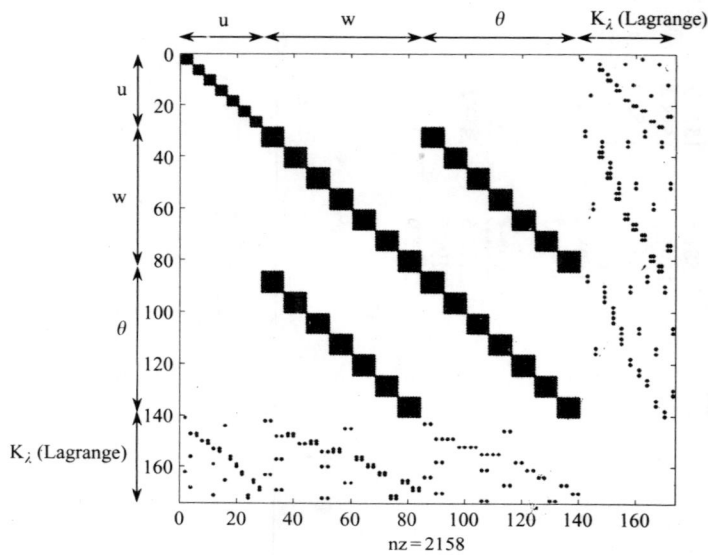

Figure 13 *Fill-in of the system matrix*

4.3.3. Constraints

Figure 13 shows the fill-in of the system matrix. Different blocks of non-zero values correspond to the local matrix K_{uu}, K_{ww}, $K_{w\theta}$ and $K_{\theta\theta}$. It also shows the coupling between the transverse deflection and the rotation. The sparse non-zero values are attached to the Lagrangian matrix for the restraints. The matrix is not positive definite.

5. Conclusions

The paper presents a coupled improved EFGM-PUM formulation to solve accurately and efficiently the vibroacoustic problem. Within the fluid, the improved EFGM, based on a defect-correction of the phase within a local basis, already gives very accurate results. With the solid, the proposed PUM formulation exhibits some bottleneck problems that must be solved: numerical quadrature and continuity for non co-linear beams. Further work will be dedicated to both aspects and to the coupling of EFGM and PUM.

Acknowledgements

The second and third authors are supported by the Région Wallonne under grant SIVA. The authors would like to thank Prof. P. Villon (Université Technogique de Compiègne) for the fruitful discussions.

6. References

[BAB 97] Babuška I., Melenk J., "The Partition of Unity method", *Int. J. Numer. Methods Eng.*, 1997, 40, pp. 727–758.

[BEL 94] Belytschko T., Lu Y. Y. and Gu L., "Element-free Galerkin Methods", *Int. J. Numer. Methods Eng.*, 1994, 37, pp. 229–256.

[BEL 96] Belytschko T., Krongauz Y., Organ D., Fleming M. and Krysl P., "Meshless Methods: An Overview and Recent Developments", *Comput. Methods Appl. Mech. Eng.*, 1996, 139, pp. 3–47.

[BOU 98a] Bouillard Ph., Ihlenburg F., "Error Estimation and Adaptivity for the Finite Element Method in Acoustics", *Advances in Adaptive Computational Methods in Mechanics*, 1998, pp. 477–492.

[BOU 98b] Bouillard Ph., Suleau S., "Element-free Galerkin Solutions for Helmholtz Problems: Formulation and Numerical Assessment of the pollution Effect", *Comput. Methods Appl. Mech. Eng.*, 1998, 162, pp. 317–335.

[CHA 97] Chadwick E., Bettes P., "Modelling of Progressive Swaves UwaveE", *Int. J. Numer. Methods Eng.*, 1997, 40, pp. 3229–3245.

[DES 98] Desmet W., *A Wave-based Prediction Technique for Coupled Vibroacoustic Analysis*, PhD Thesis nr. 98D12, KU Leuven, 1998.

[FAR 00] Farhat C., Harari I. and Franca L. P., "A Discontinuous Finite Element Method for the Helmholtz Equation", *Proceedings of the European Congress on Computational Methods in Applied Sciences and Engineering (ECCOMAS)*, 2000, Barcelona, Spain.

[GER 96] Gerdes K. and Demkowicz L., "Solution of the 3D Helmholtz Equation in Arbitrary Exterior Domains using hp-FEM and IFEM", *Comput. Methods Appl. Mech. Eng.*, 1996, 137, pp. 239–273.

[IHL 96] Ihlenburg F. and Babuška I., "Finite Element Solution of the Helmholtz Equation with High Wave Number, Part 1: The h-Version of the FEM", *Comput. Math. Applic.*, 1995, 38 (9), pp. 9–37.

[KAN 01] Kanok-Nukulchai W., Barry W. J., Saran-Yasoontorn K., Bouillard Ph., "On Elimination of Shear Locking in the Element-free Galerkin Method", *Int. J. Numer. Methods Eng.*, 2001, 52/7, pp. 705–725.

[MOE 02] Moës N. and Belytschko T., "Extended Finite Element Method for Cohesive Crack Growth", *Engineering Fracture Mechanics*, 2002, 69/7, pp. 813–833.

[MOR 95] *Fluid-Structure Interaction*, John Wiley and Sons Ltd., Chichester, England, 1995.

[URA 97] Uras R. A., Chang C. T., Chen Y. and Liu W. K., "Multiresolution Reproducing Kernel Particle Methods in Acoustics", *Journal of Computational Acoustic*, 1997, 5 (1), pp. 71–94.

[VOT 01] Voth Th. E. and Christon Mark A., "Discretization Errors Associated with Reproducing Kernel Methods: One-Dimensional Domains", *Comput. Methods Appl. Mech. Eng.*, 2001, 190 (18–19), pp. 2429–2446.

Chapter 6

A Meshfree Method for Incompressible Fluid Flows with Incorporated Surface Tension

Sudarshan Tiwari & Jörg Kuhnert
*Fraunhofer Institut Techno- und Wirtschaftmathematik Kaiserslautern,
Germany*

1. Introduction

In this paper we present a meshfree particle method for simulations of free surface flows. This is a Lagrangian method. A fluid domain is first replaced by a discrete number of points, which are referred to as 'particles'. Each particle carries all fluid information, like density, velocity, temperature etc. and moves with fluid velocity. Therefore, particles themselves can be considered as geometrical grids of the fluid domain. This method has some advantages over grid based techniques, for example, it can handle fluid domains, which change naturally, whereas grid based techniques require additional computational effort.

Numerical simulations of free surface flows have many industrial applications like casting, tank filling and others. Many methods have been developed to simulate free surface flows (Hansbo 1992, Harlow *et al.* 1965, Hirt *et al.* 1981, Kelecy *et al.* 1997, Kothe *et al.* 1992, Maronnier *et al.* 1999, Tiwari *et al.* 2000). A classical grid free Lagrangian method is *smoothed particle hydrodynamics* (SPH), which was originally introduced to solve problems in astrophysics (Lucy 1977, Gingold *et al.* 1977). It has since been extended to simulate the compressible Euler equations in fluid dynamics and applied to a wide range of problems, see (Monaghan 1992, Monaghan *et al.* 1983, Morris *et al.* 1997). The method has also been extended to simulate inviscid incompressible free surface flows (Monaghan 1994). The implementation of the boundary conditions is the main problem of the SPH method.

Another approach for solving fluid dynamic equations in a grid free framework is the moving least squares or least squares method (Belytschko *et al.* 1996, Dilts 1996, Kuhnert 1999, Kuhnert 2000, Tiwari *et al.* 2001 and 2000). With this approach boundary conditions can be implemented in a natural way just by placing the particles on boundaries and prescribing boundary conditions on them (Kuhnert 1999). The robustness of this method is shown by the simulation results in the field of airbag deployment in the car industry. Here, the membrane (or boundary) of the airbag changes very rapidly in time and takes a quite complicated shape (Kuhnert *et al.* 2000).

In (Tiwari *et al.* 2000) we have performed simulations of incompressible flows as the limit of the compressible Navier-Stokes equations with some stiff equation of state. This approach was first used in (Monaghan 1992) to simulate incompressible free surface flows by SPH. The incompressible limit is obtained by choosing a very high speed of sound in the equation of state such that the Mach number becomes small. However, the large value of the speed of sound restricts the time step to be very small due to the CFL-condition.

The projection method of Chorin (Chorin 1968) is a widely used approach to solve problems governed by the incompressible Navier-Stokes equation in a grid based structure. In (Tiwari *et al.* 2001), this method has been applied to a grid free framework with the help of the weighted least squares method. The scheme gives accurate results for incompressible Navier-Stokes equations. The occurring Poisson equation for the pressure field is solved using a grid free method. In (Tiwari *et al.* 2001), it has been shown that the Poisson equation can be solved accurately by this approach for any boundary conditions. The Poisson solver can be adopted to the weighted least squares approximation procedure with the condition that the Poisson equation and the boundary condition must be satisfied on each particle. This is a local iteration procedure.

In this paper, we further extend the scheme, presented in (Tiwari *et al.* 2001), to free surface flows. Numerical experiments are obtained with and without surface tension forces. The broken dam problem is solved without surface tension forces. Laplace's law (Landau *et al.* 1959) has been tested for different shapes of bubbles. The numerical scheme, presented here, reproduces Laplace's law exactly. Finally, the binary drop collision of liquid drops shows that the scheme is suitable for simulations of free surface flows.

The paper is organized as follows. In section 2 we present the mathematical model and boundary conditions. In section 3 the numerical scheme is described. In section 4, the weighted least squares method and its application to the *finite point-set method* (FPM) is presented. The algorithm of determination of the free surface particles is presented in section 5. Finally, some numerical tests are presented in section 6.

2. Mathematical model and boundary conditions

We consider the incompressible Navier-Stokes equations in the Lagrangian form.

$$\frac{D\vec{v}}{Dt} = -\frac{1}{\rho}\nabla p + \nu\Delta\vec{v} + \vec{g}, \qquad (1)$$

$$\nabla \cdot \vec{v} = 0. \tag{2}$$

Here, ρ is the mass density, $\vec{v}$ is the velocity vector, $\vec{g}$ is the body force acceleration vector, ν is the kinematic viscosity and p the dynamic pressure.

In addition to equations (1) and (2), appropriate initial and boundary conditions have to be provided.

For the discussion within this paper, we will consider various types of boundary conditions: solid wall, inflow, outflow, and free surface boundaries. However, we will emphasize free surface boundaries, since these are indeed a delicate problem and have many applications in industry and the sciences.

For a solid wall, one can use either free slip or no slip boundary conditions. If the viscosity is too low, a free slip condition seems to be appropriate, coupled with some model for the boundary layer. For inflow boundaries, all velocity components need to be prescribed. The surface stress boundary condition on the interface between two fluids or free surfaces is given by (Landau *et al.* 1959) as

$$[\tau \cdot \vec{n} - p\vec{n}] = \sigma \kappa \vec{n}, \tag{3}$$

where

- σ is the surface tension of the fluid, which is assumed to be constant,
- κ is the curvature on the interface,
- $\vec{n}$ is the unit normal vector on the interface, and
- τ is the viscous stress tensor is given by $\tau_{ij} = \mu \left(\partial u_i / \partial x_j + \partial u_j / \partial x_i \right)$.

The symbol $[\cdot]$ denotes the jump across the free surface boundary between two fluids. Suppose that the viscosity of one fluid adjacent to the free surface is negligible and has the pressure p_0, then the normal and tangential components of the other fluid from equation (3)

$$p - \vec{n} \cdot \tau \cdot \vec{n} = p_0 + \sigma \kappa \tag{4}$$

$$\vec{t} \cdot \tau \cdot \vec{n} = 0. \tag{5}$$

Here, $\vec{t}$ denotes the unit tangent vector on the interface.

The implementation of boundary conditions (4) and (5) requires a sufficiently good approximation of the first and second spatial derivatives of the velocity as well as of the curvature of the free surface. Please find a detailed discussion about the numerical scheme in section 3. The approximation techniques employed are described in section 4.

3. Numerical scheme

The numerical idea we present here is a generalized finite difference method for the Navier-Stokes-equations for incompressible problems with free surfaces. In fact, we call this method *finite pointset method* (FPM) for the reasons we find below.

The idea is to fill the flow domain with (numerical) points. These points are carriers of all relevant physical information (*i.e.* velocity, density, pressure etc.). In this context, we might also call these points particles, meaning they are not representing physical particles, rather they are representing a certain, finite piece of the fluid considered. The important point is that, as time evolves, the particles are moved with fluid velocity such that the numerical points (particles) move in the same fashion as a point in the fluid would move.

On the path of a numerical particle, the relevant quantities need to be updated, *i.e.* the velocity, pressure etc. will change. In order to describe these changes, we use the Navier-Stokes-equations as stated above and discretize them directly on each particle. Hence, no weak formulation is used. Since, in the Navier-Stokes equations, spatial derivatives of the velocity and the pressure appear, we will have to find a way of giving good approximations for these terms based on the knowledge of the discrete pressure and velocity values. The method employed here is the so-called weighted least squares method. Please turn to section 4 for a detailed introduction.

Summarizing, the whole idea is to fill the flow domain with numerical points (particles), each of which being a carrier of relevant physical information. We let the particles move with fluid velocity. The mean interaction radius between the particles for approximating derivatives is given by the symbol h. h is also referred to as smoothing length. As particles move in their path, physical quantities will have to be updated, governed by the Navier-Stokes-equations.

Now let us turn to the numerical method in detail. We consider the projection method described in (Chorin 1968). This is an explicit method being of first order accuracy in time. It consists of two fractional steps. At the first step we explicitly compute the new particle positions and the intermediate velocity $\vec{v}^*$ by

$$\vec{x}^{n+1} = \vec{x}^n + \delta t \vec{v}^n \tag{6}$$

$$\vec{v}^* = \vec{v}^n + \delta t \nu \Delta \vec{v}^n + \delta t \vec{g}^n \tag{7}$$

Then, at the second step, we correct $\vec{v}^*$ by solving the equation

$$\vec{v}^{n+1} = \vec{v}^* - \delta t \nabla p^{n+1} \tag{8}$$

with the incompressibility constraint

$$\nabla \cdot \vec{v}^{n+1} = 0. \tag{9}$$

Here, for simplicity, we have considered ρ to be 1. By taking the divergence of equation (8) and by making use of (9), which is the constraint that $\vec{v}^{n+1}$ must be a divergence free vector field, we come up with the Poisson equation for the pressure

$$\Delta p^{n+1} = \frac{\nabla \cdot \vec{v}^*}{\delta t} \tag{10}$$

The boundary condition for solid walls as well as for inflow boundaries is obtained by projecting equation (8) on the outward unit normal vector $\vec{n}$ to the boundary Γ. Thus, we obtain the Neumann boundary condition

$$\left(\frac{\partial p}{\partial \vec{n}} \right)^{n+1} = -\frac{1}{\delta t} \left(\vec{v}_\Gamma^{n+1} - \vec{v}_\Gamma^* \right) \cdot \vec{n},$$

where $\vec{v}_\Gamma$ is the value of $\vec{v}$ on Γ. Assuming $\vec{v} \cdot \vec{n} = 0$ on Γ, we obtain

$$\left(\frac{\partial p}{\partial \vec{n}} \right)^{n+1} = 0$$

on Γ. Moreover, the Dirichlet boundary condition

$$p = p_0 + \sigma \kappa + \vec{n} \cdot \tau \cdot \vec{n}$$

applies for a free surface as well as for outflow particles in the context of the pressure Poisson equation (10).

We note that particle positions change only in the first step. The intermediate velocity $\vec{v}^*$ is obtained for each particle on its new location. Finally, the pressure and the divergence-free velocity fields are computed also for exactly the same new particle positions.

We approximate the spatial derivatives appearing in (7) and (8) by the weighted least squares method. Furthermore, the pressure Poisson equation (10) is also solved in the least squares sense. In the following section, we describe the method of approximation of spatial derivatives. The Poisson solver, and the approximation of the curvature of free surfaces by the weighted least squares method are presented as well.

4. Weighted least squares method (WLS) and its application for FPM

In general, we would like to approximate spatial derivatives of some function. The problem is that we only know the discrete function values exactly at the particle positions. To approximate a derivative of some function at some given point, the discrete function values of the neighbor particles being in a ball about the point considered are taken into account. The WLS, which is employed for that purpose, does not require a regular grid structure. This is a big advantage for FPM.

Let $f(t, \vec{x})$ be a scalar function and $f_i(t)$ its discrete values at the particle positions $\vec{x}_i$ for $i = 1, 2, \ldots, N$ and time t. Consider the problem to approximate spatial derivatives of that particular function $f(t, \vec{x})$ at some particular particle position $\vec{x}$ based on the discrete function values of its neighbor points.

In order to restrict the number of points we introduce a weight function $w = w(\vec{x}_i - \vec{x}, h)$ with small compact support, where h determines the size of the support and represents the smoothing length. The weight function can be arbitrary; however it makes sense to choose a Gaussian weight function of the form

$$
w(\vec{x}_i - \vec{x}, h) = \begin{cases} \exp\left(-\alpha \cdot \frac{\|\vec{x}_i - \vec{x}\|^2}{h^2}\right), & \text{if } \frac{\|\vec{x}_i - \vec{x}\|}{h} \leq 1 \\ 0, & \text{else,} \end{cases}
$$

where α is a positive constant and is considered to be in the range of 6. The size of h defines a set of neighbor particles around $\vec{x}$. So far, in our implementation, we allow user given h as a function in space and time. However, no adaptive choice of h is realized yet. Working with user given h implies that new particles will have to be brought into play as the particle distribution becomes too sparse or, logically, particles will have to be removed from the computation domain as they become too dense.

Let $P(\vec{x}, h) = \{\vec{x}_i : i = 1, 2, \ldots, n\}$ be the set of n neighbor points of $\vec{x}$ in a ball of radius h. For consistency reasons, some obvious restrictions are required, for example, in 2D there should be at least 5 neighbor particles and they should neither be on the same line nor on the same circle.

The determination of derivatives of a function can be computed easily and accurately by using the Taylor series expansion and the least squares approximation. We write Taylor's expansion about the point $\vec{x}$ with unknown coefficients and then compute these coefficients by minimizing a weighted error over the neighbor points.

Hence, consider Taylor's expansion of $f(t, \vec{x}_i)$ about $\vec{x}$

$$
f(t, \vec{x}_i) = f(t, x) + \sum_{k=1}^{3} f_k(t, \vec{x})\left(x_i^{(k)} - x^{(k)}\right)
$$

$$
+ \frac{1}{2}\sum_{k,l=1}^{3} f_{kl}(t, \vec{x})\left(x_i^{(k)} - x^{(k)}\right)\left(x_i^{(l)} - x^{(l)}\right) + e_i,
$$

where e_i is the error in Taylor's expansion at the point $\vec{x}_i$. The symbol $\vec{x}_i^{(k)}$, represents the k-th component of the particle position $\vec{x}_i$. The unknowns f_k and $f_{kl}(= f_{lk})$ for $k, l = 1, 2, 3$ represent the approximations of the first and second derivatives of f and are computed by minimizing the error e_i for $i = 1, 2, \ldots, n$, where $f(t, \vec{x}) = f$ is the known discrete function value at the particle position $\vec{x}$. The system of equations can be written as

$$\vec{e} = M\vec{a} - \vec{b},$$

where

$$M = \begin{pmatrix} \Delta x1_1 & \Delta x2_1 & \Delta x3_1 & \Delta x11_1 & \Delta x12_1 & \Delta x13_1 & \Delta x22_1 & \Delta x23_1 & \Delta x33_1 \\ \Delta x1_2 & \Delta x2_2 & \Delta x3_2 & \Delta x11_2 & \Delta x12_2 & \Delta x13_2 & \Delta x22_2 & \Delta x23_1 & \Delta x33_1 \\ \vdots & \vdots & \vdots & \vdots & \vdots & \vdots & \vdots & \vdots & \vdots \\ \Delta x1_n & \Delta x2_n & \Delta x3_n & \Delta x11_n & \Delta x12_n & \Delta x13_n & \Delta x22_n & \Delta x23_n & \Delta x33_n \end{pmatrix},$$

$$\vec{a} = [f_1, f_2, f_3, f_{11}, f_{12}, f_{13}, f_{22}, f_{23}, f_{33}]^t,$$

$$\vec{b} = [f_1 - f, f_2 - f, \ldots, f_n - f]^t \quad \text{and} \quad \vec{e} = [e_1, e_2, \ldots, e_n]^t.$$

The symbols are defined as

$$\Delta xk_i = x_i^{(k)} - x^{(k)}, \quad \Delta xkl_i = (x_i^{(k)} - x^{(k)})(x_i^{(l)} - x^{(l)}) \quad \text{with } (k \neq l)$$

and

$$\Delta xkk_i = \frac{1}{2}(x_i^{(k)} - x^{(k)})^2 \quad \text{for } k, l = 1, 2, 3 \text{ and } i = 1, 2, \ldots, n.$$

For $n \geq 10$, this system is over-determined for the nine unknowns f_k and $f_{kl}(= f_{lk})$ for $k, l = 1, 2, 3$.

The unknowns in the vector $\vec{a}$ are obtained from a weighted least squares method by minimizing the quadratic form

$$J = \sum_{i=1}^{n} w_i e_i^2.$$

The above equations can be expressed as

$$J = (M\vec{a} - \vec{b})^t W (M\vec{a} - \vec{b})^{-1},$$

with

$$W = \begin{pmatrix} w_1 & 0 & \cdots & 0 \\ 0 & w_2 & \cdots & 0 \\ \vdots & \vdots & \ddots & \vdots \\ 0 & 0 & \cdots & w_n \end{pmatrix},$$

where $w_i = w(\vec{x}_i - \vec{x}, h)$. The minimization of J formally yields

$$\vec{a} = (M^t W M)^{-1} (M^t W) \vec{b} \tag{11}$$

The Taylor expansion may include high order expansion. The employment of particular weight functions can force the least square approximations to recover the finite difference discretization in the special case that all particles are placed in a regular grid structure.

4.1. Weighted least squares approach for the Poisson equation

As we have seen in the description of the numerical scheme in section 3, we need to solve the pressure Poisson equation

$$\Delta p = \frac{\nabla \cdot \vec{v}^*}{\delta t} \tag{12}$$

with the boundary conditions

$$\frac{\partial p}{\partial \vec{n}} = 0$$

for solid walls as well as for inflow boundaries and

$$p = p_0 + \sigma \kappa + \vec{n} \cdot \tau \cdot \vec{n}$$

for free surface as well as for outflow boundaries. Here, the symbol p denotes p^{n+1} for the sake of simplicity.

Since we are in a grid-free structure, it is not obvious to apply classical methods like finite difference or finite element methods for a numerical scheme solving the above Poisson equation. Of course, one could construct a regular grid and solve the Poisson equation by some classical finite difference method and then interpolate the results of pressure back to the original particle distribution. However, this will give smearing effects and is possibly be of high computational effort especially if geometries become complex.

Therefore, we use a local iteration approach on the basis of the least squares approximation, which the Poisson equation is forced to strictly satisfy. The main advantage is that this procedure can be applied directly to the given particle distribution. This method is stable and gives accurate results for all boundary value problems of the Poisson equation, see (Tiwari et al. 2001) for details.

In the beginning of this section, we have presented the least squares method to approximate derivatives of a function at an arbitrary point from its neighbor values. Now we have a slightly different situation. Based on the function values of the neighbor particles, we would like to compute an approximate function value under the condition that some determined value of the approximate Laplacian is fulfilled. The pressure values at the new particle positions are not yet known. Therefore, the least squares approach cannot be applied directly as described in the previous subsection. Hence, we prescribe an initial guess $p^{(0)}$ for the pressure p. Now, we consider the problem of determining p at an arbitrary particle position $\vec{x}$ from its neighbor points $\vec{x}_i, i = 1, \ldots, n$. As in the previous section, we again consider a Taylor expansion of p about some point $\vec{x}$

$$p^{(j)}(\vec{x}_i) = p^{(j+1)}(\vec{x}) + \sum_{k=1}^{3} p_k^{(j+1)}(\vec{x}) \left(x_i^{(k)} - x^{(k)} \right)$$

$$+ \frac{1}{2} \sum_{k,l=1}^{3} p_{kl}^{(j+1)}(\vec{x}) \left(x_i^{(k)} - x^{(k)} \right) \left(x_i^{(l)} - x^{(l)} \right) + e_i^{(j+1)} \qquad (13)$$

for $j = 0, 1, 2, \ldots$, where $p^{(0)}(\vec{x}_i)$ are the given initial discrete particle values. We require that the Poisson equation (12) be satisfied at $\vec{x}$. Hence, we have to add the following equation to the set of n equations in (13)

$$\frac{\nabla \cdot \vec{v}^*}{\delta t} = p_{11}^{(j+1)}(\vec{x}) + p_{22}^{(j+1)}(\vec{x}) + p_{33}^{(j+1)}(\vec{x}).$$

If $\vec{x}$ is a particle of some solid wall or inflow boundary, we also have to enforce the Neumann boundary condition to strictly satisfy by adding the equation

$$0 = p_1^{(j+1)}(\vec{x})n_x + p_2^{(j+1)}(\vec{x})n_y + p_3^{(j+1)}(\vec{x})n_z$$

to the given system of $n + 1$ equations. Here, n_x, n_y, n_z are the respective components of the unit normal vector $\vec{n}$.

If the particle belongs to a free surface or outflow boundary, we have the Dirichlet condition satisfied strictly by adding the equation

$$p_\Gamma(\vec{x}) = p^{(j+1)}(\vec{x}).$$

Here, $p_\Gamma(\vec{x})$ is a user given boundary value for the pressure. Summing up, for boundary particles, we have a total of $n + 2$ equations for 10 unknowns. In general, the number of neighbors is greater than 10.

The coefficients we obtain by minimizing the residuals e_i are

$$p^{(j+1)}, p_1^{(j+1)}, p_2^{(j+1)}, p_3^{(j+1)}, p_{11}^{(j+1)}, p_{12}^{(j+1)}, p_{13}^{(j+1)}, p_{22}^{(j+1)}, p_{23}^{(j+1)}, p_{33}^{(j+1)}$$

for $j = 0, 1, 2, \ldots$ at a particular location $\vec{x}$. For example, the functional to be minimized for a boundary particle with Neumann boundary condition reads as

$$J = \sum_{i=1}^{n} w_i \left(e_i^{(j+1)} \right)^2 + \left(\Delta p^{(j+1)} - \frac{\nabla \cdot \vec{v}^*}{\delta t} \right)^2 + \left(\frac{\partial p^{(j+1)}}{\partial \vec{n}} - 0 \right)^2.$$

Similarly to (11), the minimization of J is given by

$$\vec{a}^{(j+1)} = (M^t W M)^{-1} (M^t W) \vec{b}^{(j)}, \quad j = 0, 1, 2, \ldots,$$

where the matrices and the vectors differ slightly from (11) and are given by

$$M = \begin{pmatrix} 1 & \Delta x 1_1 & \Delta x 2_1 & \Delta x 3_1 & \Delta x 11_1 & \Delta x 12_1 & \Delta x 13_1 & \Delta x 22_1 & \Delta x 23_1 & \Delta x 33_1 \\ \vdots & \vdots & \vdots & \vdots & \vdots & \vdots & \vdots & \vdots & \vdots & \vdots \\ 1 & \Delta x 1_n & \Delta x 2_n & \Delta x 3_n & \Delta x 11_n & \Delta x 12_n & \Delta x 13_n & \Delta x 22_n & \Delta x 23_n & \Delta x 33_n \\ 0 & 0 & 0 & 0 & 1 & 0 & 0 & 1 & 0 & 1 \\ 0 & n_x & n_y & n_z & 0 & 0 & 0 & 0 & 0 & 0 \end{pmatrix}.$$

$$W = \begin{pmatrix} w_1 & 0 & \cdots & \cdots & \cdots & 0 & 0 & 0 \\ 0 & w_2 & \cdots & \cdots & \cdots & 0 & 0 & 0 \\ \vdots & \vdots & \ddots & \ddots & \ddots & \vdots & \vdots & \vdots \\ 0 & 0 & \cdots & \cdots & \cdots & w_n & 0 & 0 \\ 0 & 0 & 0 & 0 & 0 & 0 & 1 & 0 \\ 0 & 0 & 0 & 0 & 0 & 0 & 0 & 1 \end{pmatrix},$$

$$\vec{a}^{(j+1)} = \left[p^{(j+1)}, p_1^{(j+1)}, p_2^{(j+1)}, p_3^{(j+1)}, p_{11}^{(j+1)}, p_{12}^{(j+1)}, p_{13}^{(j+1)}, p_{22}^{(j+1)}, p_{23}^{(j+1)}, p_{33}^{(j+1)} \right],$$

$$\vec{b}^{(j)} = \left[p_1^{(j)}, p_2^{(j)}, \ldots, p_n^{(j)}, \frac{\nabla \cdot \vec{v}^*}{\delta t}, 0 \right]^t.$$

The scheme (13) is clearly an iterative process. The iteration is stopped if the local error satisfies

$$\frac{\sum_{i=1}^{N} \left| p_i^{(j+1)} - p_i^{(j)} \right|}{\sum_{i=1}^{N} \left| p_i^{(j+1)} \right|} \leq \varepsilon.$$

Finally, the solution is defined by $p(\vec{x}_i) := p^{(j+1)}(\vec{x}_i)$ as j tends to infinity. The parameter ε is a very small positive constant and can differ according to the size of h. The convergence rate is faster if h is taken larger. Therefore, multigrid approaches can indeed be useful in order to reduce the computational effort.

·Of course, it is necessary to prescribe the initial value of the pressure at time $t = 0$. For the pressure iteration, the initial guess of the pressure for time level $n + 1$ is taken as the pressure from time level n.

4.2. *Approximation of derivatives of velocities on the free surface*

For free surface particles we are required to include the boundary conditions (4) and (5) into the approximation of spatial derivatives in (7). For the sake of simplicity we consider the case of two spatial dimensions. The boundary conditions (4) and (5) can be explicitly written as

$$2\mu \left(\frac{\partial u}{\partial x} n_x^2 + \frac{\partial u}{\partial y} n_x n_y + \frac{\partial v}{\partial x} n_x n_y + \frac{\partial v}{\partial y} n_y^2 \right) = p - p_0 + \sigma \kappa \qquad (14)$$

$$2 \frac{\partial u}{\partial x} n_x n_y + \frac{\partial u}{\partial y} (n_y^2 - n_x^2) + \frac{\partial v}{\partial x} (n_y^2 - n_x^2) + 2 \frac{\partial v}{\partial y} n_x n_y = 0. \qquad (15)$$

Here, u, v denote the respective components of the velocity vector $\vec{v}$ and n_x, n_y denote the components of the unit normal vector and the tangent vector is defined by $\vec{t} = (-n_y, n_x)$.

The incorporation of (14) and (15) into the approximation of derivatives of velocities on the free surface is not straightforward. (14) and (15) both contain the first derivatives of both u and v. On the other hand, for example, the derivatives of u are obtained by Taylor's expansion, which contains only the derivatives of u but not of v. Therefore, we have to compute derivatives of u and v together. The method is again an extension of the occurring least squares matrix. Suppose that we want to approximate the derivatives of u and v on the free surface particle located at (x, y) from its neighbor values. Let $(u, v) = (u, v)(x, y)$, $(u, v)_i = (u, v)(x_i, y_i)$, $dx_i = x_i - x, dy_i = y_i - y$ for $i = 1, \ldots, n$. Consider Taylor's expansion of u and v around (x, y)

$$u_i = u + \frac{\partial u}{\partial x} dx_i + \frac{\partial u}{\partial y} dy_i + \frac{1}{2} \frac{\partial^2 u}{\partial x^2} dx_i^2 + \frac{\partial^2 u}{\partial x \partial y} dx_i dy_i + \frac{1}{2} \frac{\partial^2 u}{\partial y^2} dy_i^2 + e_{ui}$$

$$(16)$$

and

$$v_i = v + \frac{\partial v}{\partial x} dx_i + \frac{\partial v}{\partial y} dy_i + \frac{1}{2} \frac{\partial^2 v}{\partial x^2} dx_i^2 + \frac{\partial^2 v}{\partial x \partial y} dx_i dy_i + \frac{1}{2} \frac{\partial^2 v}{\partial y^2} dy_i^2 + e_{vi}$$

$$(17)$$

As in the previous cases, we can rewrite equations (16) and (17) in matrix form as

$$\vec{e} = M\vec{a} - \vec{b},$$

where

$$
M = \begin{pmatrix}
dx_1 & dy_1 & \frac{1}{2}dx_1^2 & dx_1 dy_1 & \frac{1}{2}dy_1^2 & 0 & 0 & 0 & 0 & 0 \\
\vdots & \vdots & \vdots & \vdots & \vdots & \vdots & \vdots & \vdots & \vdots & \vdots \\
dx_n & dy_n & \frac{1}{2}dx_n^2 & dx_n dy_n & \frac{1}{2}dy_n^2 & 0 & 0 & 0 & 0 & 0 \\
0 & 0 & 0 & 0 & 0 & dx_1 & dy_1 & \frac{1}{2}dx_1^2 & dx_1 dy_1 & \frac{1}{2}dy_1^2 \\
\vdots & \vdots & \vdots & \vdots & \vdots & \vdots & \vdots & \vdots & \vdots & \vdots \\
0 & 0 & 0 & 0 & 0 & dx_n & dy_n & \frac{1}{2}dx_n^2 & dx_n dy_n & \frac{1}{2}dy_n^2
\end{pmatrix}
$$

$$(18)$$

$$\vec{a} = [u_x, u_y, u_{xx}, u_{xy}, u_{yy}, v_x, v_y, v_{xx}, v_{xy}, v_{yy}]^t$$

and

$$\vec{b} = [u_1 - u, \ldots, u_n - u, v_1 - v, \ldots, v_n - v]^t.$$

The above system has $2n$ equations with 10 unknowns. The minimization of the error gives the derivatives of both velocity components together. However, one has to invert a 10-by-10 matrix instead of a 5-by-5 matrix. Therefore, we use this larger system only for the few free surface particles. For the incorporation of the boundary conditions (14) and (15), we have to add these equations to the system (16), (17), where the matrix (18) is enhanced by two lines in the sense

$$
\begin{matrix}
2\mu n_x^2 & 2\mu n_x n_y & 0 & 0 & 0 & 2\mu n_x n_y & 2\mu n_y^2 & 0 & 0 & 0 \\
2n_x n_y & (n_y^2 - n_x^2) & 0 & 0 & 0 & (n_y^2 - n_x^2) & 2n_x n_y & 0 & 0 & 0
\end{matrix}
$$

and the right hand side vector $\vec{b}$ is given by

$$\vec{b} = [u_1 - u, \ldots, u_n - u, v_1 - v, \ldots, v_n - v, p - p_0 + \sigma \kappa, 0]^t.$$

The minimization process is the same as above.

4.3. Approximation of the local curvature on free surfaces

Boundary condition (4) requires the knowledge of the curvature of the free surface. In this section, we describe the approximation of the curvature in the two dimensional case. We approximate a circle of radius R with the center (x_c, y_c) running through the free surface particle located at (x, y) such that it fits, locally, all the neighbor-surface-points in a least squares sense. The curvature κ and the unit normal vector on free surface at (x, y) are given by

$$\kappa = \frac{1}{R},$$

$$n_x = (n - x_c)\kappa, \quad n_y = (y - y_c)\kappa.$$

If the center of the circle lies outside of the fluid considered, the sign of curvature is taken negative.

A circle is represented by a general second order equation

$$x^2 + y^2 + Dx + Ey + F = 0,$$

where D, E, F are to be determined. The radius and the center of the circle running through (x, y) is given by

$$x_c = -\frac{D}{2}, \quad y_c = -\frac{E}{2}, \quad R = \frac{1}{2}\sqrt{D^2 + E^2 - 4F}, \quad (D^2 + E^2 > 4F).$$

Now, we have to determine the coefficients D, E, F at every free surface particle (x, y) from its free surface neighbor particles $(x_i, y_i), i = 1, \ldots, n$. Here (x, y) is one of the (x_i, y_i). In general, there are more than three neighbor points, therefore these coefficients are approximated by the weighted least squares method, as described above. In order to avoid the least squares approximation, one can choose two nearest neighbors (one left and one right) of (x, y) such that the circle can be fitted more accurately. Singularities may occur, if all free surface particles lie on the same straight line. In this case the curvature is considered to be zero.

5. Determination of the free surface particles

In this section we would like to give a brief description of the strategy of how to indicate particles belonging to a free surface. We would like to remind the reader that these particles are not known *a priori*; however, it is important to have a very accurate selection of them, otherwise the whole numerical procedure and application of boundary conditions is likely to fail. For the determination of the free surface particles, we come up with a definition. We say that a particle at the position $\vec{x}_i$ belongs to some free surface, if we can place a sphere in the neighborhood of the particle such that

(i) $\vec{x}_i$ belongs to the surface of the sphere (*i.e.* it is not the center)
(ii) the radius of the sphere is $r_S = \alpha h$ where h is the smoothing length and α is a constant, preferably in the range $\alpha \in [0.7, 1.0]$
(iii) no other particle lies inside the sphere.

This definition is rather theoretical, but it makes sense. If a particle is really at the free surface, then there will be indeed such a sphere, because one half-space is more or less empty for surface particles. An interior particle, however, should not find such a sphere, or, in other words, if it would find a sphere meeting the above conditions, then this would mean there is a big hole in the interior of the flow domain, and this is not acceptable from the point of view of computational accuracy. Consequently, this means that interior holes have to be stuffed with particles before their radius tends to reach the magnitude of r_S.

Obeying these rules, we have a unique description of particles at the free surface. More problematic is the implementation of the whole idea. To search for appropriate holes for one particle (for instance for the particle at position $\vec{x}_i$), it takes about $25 M^2$ floating point operations, where M is the number of relevant neighbor particles related to the position $\vec{x}_i$. However, M is usually in the range of $M \in [20, 50]$ for 2D-applications and $M \in [40, 90]$ for 3D-applications, depending on the particle configuration. Hence, the effort of searching surface particles is huge and can take 10 percent of the overall computation time. The idea to reduce that effort is to

(i)   consider only those particles as candidates for being at the free surface at time level t_n if they were in the neighborhood of a free surface particle at time level t_{n-1} (this reduces the number of particles to be checked)
(ii) do the search for the free surface particles not for each time step.

Both methods mentioned above have shown excellent applicability.

6. Numerical tests

6.1. *Breaking dam problem*

The breaking dam problem is a very popular and simple test case, which helps to validate numerical schemes for the simulation of free surface flows. It consists of a simple initial configuration and simple initial and boundary conditions. In (Martin *et al.* 1952) the experimental results are reported and several authors have reported their numerical results (Hansbo 1992, Hirt *et al.* 1981, Kelecy *et al.* 1997, Maronnier *et al.* 1999, Monaghan 1994).

Consider a rectangular column of water with a width of $a = 0.1$ m and a height of 0.2 m. The lines $x = 0, y = 0$, and $x = 0.6$ consist of the solid wall. In the simulation, the upper and the right boundary of the water columns are considered as the free surface boundary. Initially, 1136 particles are distributed randomly. The size of the smoothing length is $h = 0.01$. The gravity is $g = 9.81$ m/s^2 and acts downwards. The initial velocity is set at zero. The initial pressure p_0 is also considered to be zero. The air pressure is assumed to be zero and surface tension forces are neglected. When the right wall (dam) is removed, the water column

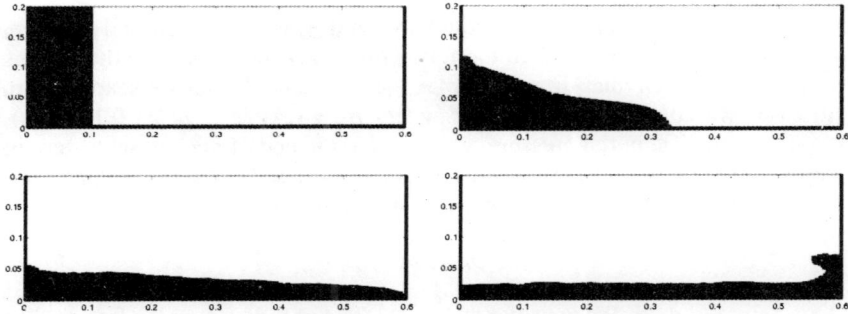

Figure 1 *Particle positions at successive times*

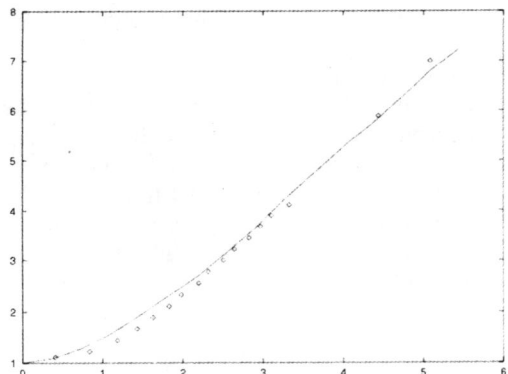

Figure 2 *Dimensionless front position x/a versus dimensionless time $t/\sqrt{2g/a}$*

collapses under the influence of gravity. The density and the viscosity of the fluid are $\rho = 1\,\text{kg/m}^3$, $\mu = 0.0004\,\text{kg/(ms)}$. No slip boundary condition is used on the solid walls. The particles, plotted successively in time, are shown in Figure 1.

In Figure 2 the position of the leading fluid front versus time is compared with experimental results (Martin *et al.* 1952). The figure shows a good agreement between the numerical and experimental results.

6.2. *Laplace's law*

It is well known that a drop of arbitrary shape becomes spherical due to the surface tension forces on the free surface boundary. In the equilibrium state a bubble should satisfy the Laplace law

$$p_l - p_g = \sigma\kappa.$$

Here, p_l is the pressure inside of the liquid drop and p_g is the background pressure, which is considered to be zero. In the following, we have considered three types of drops, the exact circular drop, the octagonal drop and the square shaped drop. In all cases we consider the fluid parameters $\rho = 1\,\text{kg/m}^3$, $\mu = 0.1\,\text{kg(ms)}$, $\sigma = 1$ dynes/m. The initial pressure, velocity and the body force are set to zero in all cases. The drop pressure is considered as the average pressure of all particles. Hence, in the equilibrium, the following relation must hold

$$p_l = \sigma\kappa.$$

(a) Exact circular drop: We consider the exact circle of radius 1 on the free surface and we generate particles inside as shown in Figure 3. The particles are placed at the distance of 1/10. In this case we obtain a curvature of 1 along the free surface boundary. The initial pressure of the drop is considered to be zero. After time $t = 0.002$ s the pressure reaches 1. The maximum velocity is 2.9952e-7.

(b) Octagonal drop: Here, we approximate a circle by an octagon. In this case there are 8 corner points, where the curvatures naturally are higher. Many authors (Lafaurie *et al*. 1994, Ginzburg *et al*. 2001) reported the "spurious" or "anomalous" currents around the free surface. In practice we always obtain a n-gon for finer grids. Hence, the large curvatures on the corners of the free surface boundary produce such currents as shown in Figure 4. The drop reaches equilibrium and becomes circular in the steady state, see Figure 5. In the steady state, the maximum velocity is equal to 3.88069e-4 and the drop pressure is equal to 1.06943992. Figure 6

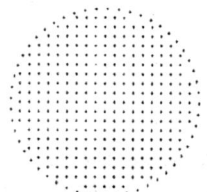

Figure 3 *Exact circular drop*

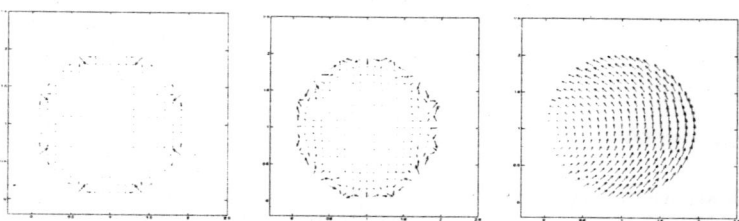

Figure 4 *Velocity profile of octagonal drop at time $t = 0.001$ s (left), $t = 0.301$ s (middle) and $t = 5.001$ s (right)*

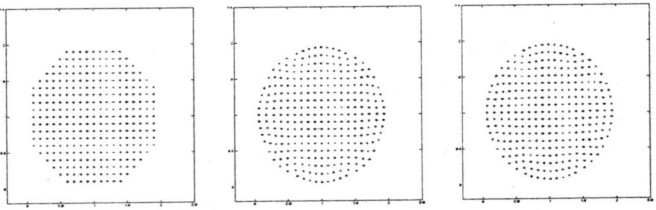

Figure 5 *Positions of particles octagonal drop at time t = 0.0 s (left), t = 0.301 s (middle) and t = 5.001 s (right)*

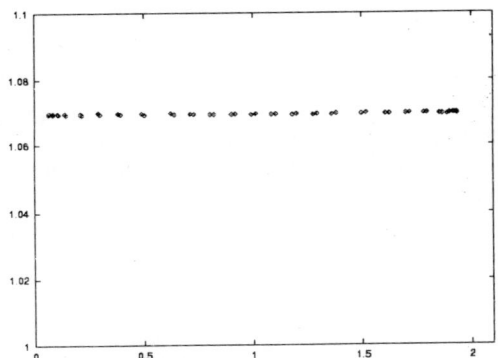

Figure 6 *The curvature on free surface of the octagonal drop at t = 5.001 s*

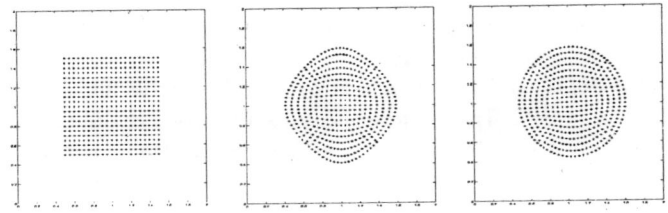

Figure 7 *Square drop at time t = 0.001 s (left), t = 0.301 s (middle) and t = 5.001 s (right)*

shows that the relation between surface curvature and drop pressure justifies the Laplace law.

(c) Square shaped drop: As a final test of the Laplace law, we consider a square shaped drop. In this case there are four corners representing naturally large curvature. The value of the curvatures on the other free surface particles is zero. In Figure 7, we have plotted the time evolution of the drop. After a short time, it shows some oscillating behavior but finally it reaches a state of equilibrium. For

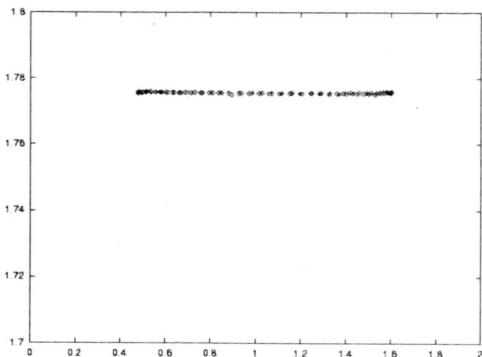

Figure 8 *The curvature on free surface of the square drop at* $t = 5.001$ s

larger viscosity, the drop reaches equilibrium without oscillation. In the equilibrium state the maximum velocity is equal to 1.47952134e-3 and the pressure is equal to 1.77552119. Hence, the value of the pressure and the curvature of the drop in equilibrium justify the Laplace law (see Figure 8).

6.3. *Drop collisions*

We consider two drops of the same size moving with the same magnitude of velocities in opposite directions. The magnitude of the initial velocity is 4 m/s. The body force of the drops is considered to be zero. The radii of the drops equal to 1 m and the initial spacing of the particles is 1/10. The density and viscosity are $\rho = 1$ kg/m^3 and $\mu = 1$ kg(ms). Hence the Reynolds number is Re $= \rho U D / \mu = 16$, where $U = 8$ m/s is the relative velocity, $D = 2$ is the diameter of drop. The surface tension coefficient is set at $\sigma = 1$ dynes/m, such that the Weber number becomes $We = \rho D U^2 / \sigma = 128$. The numerical results are very close to the experimental and other numerical results, presented in (Ash *et al.* 1990, Kothe *et al.* 1992, Lafaurie *et al.* 1994).

Two types of collisions are considered. The first is the head-on collision. In the parameters mentioned above both drops are merging into a single drop after collision. As we see in Figure 9, the drop becomes elliptical at time $t = 0.301$ s. Due to the higher surface tension on the top and bottom of the drop, it starts to shrink the top and bottom and stretching the left and right side. The results are comparable with those presented in (Ash *et al.* 1990). Finally, it reaches the equilibrium state and the shape remains unchanged. We simply wanted to test whether the presented method works for simulations of free surface flows and therefore have not tested the collisions for higher surface tension force and larger Reynolds number.

As a second example of drop collision we consider the non-central collision with impact parameter $B = 0.25$. Other input parameters are same as in the case

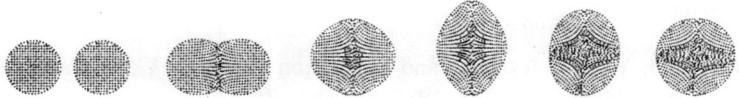

Figure 9 *Head on collision at time $t = 0.001$ s, $t = 0.101$ s, $t = 0.201$ s, $t = 0.301$ s, $t = 3.001$ s and $t = 5.001$ s (from left to right)*

Figure 10 *Non-central collision with impact parameter 0.25 collision at time $t = 0.001$ s, $t = 0.101$ s, $t = 0.201$ s, $t = 0.301$ s, $t = 0.701$ s, $t = 1.251$ s, $t = 3.001$ s and $t = 5.001$ s (from left to right and top to bottom)*

of head on collision. In contrast to the first case, the drop is rotating after collision. Shrinking and stretching of drop due to the effect of surface tension is similar to the head-on collision. The time evolution of the drop is presented in Figure 10.

7. Conclusion

A meshfree method is used to simulate free surface flows. The incompressible Navier-Stokes equations are used as a mathematical model. The numerical experiments are performed with and without surface tension force on free surface. The spatial derivatives of the Navier-Stokes equations are approximated by the weighted least squares method. The pressure Poisson equation is solved by the least squares method. Free surface boundary conditions can be directly included in the least squares approximation. Locations of free surface particles are determined by a very simple approach. Close agreements between numerical and experimental results show the robustness of the scheme. Future work will be the extension of the method in 3D.

Acknowledgments

The authors would like to thank Deutsche Forschungsgemeinschaft for their financial support.

8. References

Ash N., Poo J. Y., "Coalescence and Separation in Binary Collisions of Liquid Drops", *J. Fluid Mech.*, Vol. 221, 1990, pp. 183–204.

Belytschko T., Krongauz Y., Flemming M., "Organ D., Liu W.K.S., Smoothing and Accelerated Computations in the Element-free Galerkin Method", *J. Comp. Appl. Maths.*, Vol. 74, 1996, pp. 111–126.

Chorin A., "Numerical Solution of the Navier-Stokes Equations", *J. Math. Comput.*, Vol. 22, 1968, pp. 745–762.

Dilts G. A., "Moving Least Squares Particle Hydrodynamics. I: Consistency and Stability", *Hydrodynamics methods group report*, Los Alamos National Laboratory, 1996.

Gingold R. A., Monaghan J. J., "Smoothed Particle Hydrodynamics: Theory and Application to Non-spherical Stars", *Mon. Not. Roy. Astron. Soc.*, Vol. 181, 1977, pp. 375–389.

Ginzburg I., Wittum G., "Two-phase Flows on Interface Refined Grids Modeled with VOF, Staggered Finite Volumes, and Spline Iterpolants", *J. Comp. Phys.*, Vol. 166, 2001, pp. 302–335.

Hansbo P., "The Characteristic Streamline Diffusion Method for the Time-dependent Incompressible Navier-Stokes Equations", *Comp. Meth. Appl. Mech. Eng.*, Vol. 99, 1992, pp. 171–186.

Harlow F. H., Welch J. E., "Numerical Study of Large Amplitude Free Surface Motions", *Phys. Fluids*, Vol. 8, 1965, p. 2182.

Hirt C. W., Nichols B. D., "Volume of Fluid (VOF) Method for Dynamic of Free Boundaries", *J. Comput. Phys.*, Vol. 39, 1981, p. 201.

Kelecy F. J., Pletcher R. H., "The Development of Free Surface Capturing Approach for Multi-dimensional Free Surface Flows in Closed Containers", *J. Comput. Phys.*, Vol. 138, 1997, p. 939.

Kothe D. B., Mjolsness R. C., "RIPPLE: A New Model for Incompressible Flows with Free Surfaces", *AIAA Journal*, Vol. 30, No 11, 1992, pp. 2694–2700.

Kuhnert J., *General Smoothed Particle Hydrodynamics*, Ph.D. Thesis, Kaiserslautern University, Germany, 1999.

Kuhnert J., "An Upwind Finite Pointset Method for Compressible Euler and Navier-Stokes Equations", preprint, ITWM, Kaiserslautern, Germany, 2000.

Kuhnert J., Tramecon A., Ullrich P., "Advanced Air Bag Fluid Structure Coupled Simulations applied to out-of Position Cases", *EUROPAM Conference Proceedings 2000*, ESI group, Paris, France.

Landau L. D., Lifshitz E. M., *Fluid Mechanics*, Pergamon, New York, 1959.

Lafaurie B., Nardone C., Scardovelli R., Zaleski S., Zanetti G., "Modelling Merging and Fragmentation in Multiphase Flows with SURFER", *J. Comput. Phys.*, Vol. 113, 1994, pp. 134–147.

Lucy L. B., "A Numerical Approach to the Testing of the Fission Hypothesis", *Astron, J.*, Vol. 82, 1977, p. 1013.

Maronnier V., Picasso M., Rappaz J., "Numerical Simulation of Free Surface Flows", *J. Comput. Phys.* Vol. 155, 1999, p. 439.

Martin J. C., Moyce M. J., "An Experimental Study of the Collapse of Liquid Columns on a Liquid Horizontal Plate", *Philos. Trans. Roy. Soc. London*, Ser. A 244, 1952, p. 312.

Monaghan J. J., "Smoothed Particle Hydrodynamics", *Annu. Rev. Astron. Astrop.*, Vol. 30, 1992, pp. 543–574.

Monaghan J. J., "Simulating Free Surface Flows with SPH", *J. Comput. Phys.*, Vol. 110, 1994, p. 399.

Monaghan J. J., Gingold R. A., "Shock Simulation by Particle Method SPH", *J. Comp. Phys.*, Vol. 52, 1983, pp. 374–389.

Morris J. P., Fox P. J., Zhu Y., "Modeling Low Reynolds Number Incompressible Flows Using SPH", *J. Comput. Phys.*, Vol. 136, 1997, pp. 214–226.

Tiwari S., Kuhnert J., "Grid Free Method for Solving Poisson Equation", *Berichte des Fraunhofer ITWM*, Kaiserslautern, Germany, Nr. 25, 2001.

Tiwari S., Kuhnert J., "Finite Pointset Method Based on the Projection Method for Simulations of the Incompressible Navier-Stokes Equations", to appear in M. Griebel, M. A. Schweitzer (Eds.), Springer LNCSE: *Meshfree Methods for Partial Differential Equations*, Springer-Verlag, 2003.

Tiwari S., Kuhnert J., *Particle Method for Simulations of Free Surface Flows*, preprint Fraunhofer ITWM, Kaiserslautern, Germany, 2000.

Tiwari S., "A LSQ-SPH Approach for Compressible Viscous Flows", to appear in *Proceedings of the 8th International Conference on Hyperbolic Problems Hyp2000*.

Tiwari S., Manservisi S., *Modeling Incompressible Navier-Stokes Flows by LSQ-SPH*, Berichte des Fraunhofer ITWM, Kaiserslautern, Germany, 2000.

Chapter 7

On the Numerical Solution of Unsteady Fluid Flow Problems by a Meshless Method

Tonino Sophy & Hamou Sadat
Laboratoire d'Études Thermiques (UMR 6608) ESIP, Poitiers, France

1. Introduction

In spite of the great success of the finite element method as an effective numerical method for the solution of partial differential equations on complex domains, there has been a growing interest in meshless methods over recent years [ALU 01, BEL 94, CHA 01, LIU 01, OSH 01, ZHA 00]. For our part, we have developed a diffuse approximation based collocation method for solving incompressible steady fluid flows [SAD 95, SAD 96]. One of the primary issues in these problems, whether a regular or unstructured type grid is used, is how to handle the pressure-velocity coupling. This is an important issue since an explicit equation for the pressure does not exist. The pressure-velocity coupling problem can be avoided by using a streamfunction-vorticity approach. In this case, we have shown that the diffuse approximation method is as accurate as the well known control-volume based finite element method [PRA 98]. However streamfunction-vorticity methods are not readily extended to three dimensions. Therefore, the method has been extended to the primitive variables formulation of the Navier-Stokes equations by means of a projection algorithm [COU 98]. The Poisson equation that arises from the pressure correction process consumes, however, a large portion of the computational time.

Since the memory resource in many cases is limited for large-scale problems, direct methods are seldom used and iterative schemes are preferred. The conjugate gradient algorithm is a very powerful method for solving symmetric positive definite sparse linear systems, especially when it is used with a preconditioner. In this algorithm, the residual vector is minimized in each iteration step with respect to some suitable norm. During this process, the residual vectors are constructed in such a way that they are orthogonal to each other with regard to the Euclidian inner product. Additionally, because of the symmetry of the matrix, the residual vectors fulfill a three-term recursion, which is a characteristic of the algorithm. However, this algorithm fails in general for nonsymmetric or indefinite linear systems. Several attempts have then been made to come up with a generalization of this method for the nonsymmetric case. One can, for example, maintain the minimization property by choosing the direction vector as a linear combination of the residual vector and previous direction vectors. This approach has been used in methods like Orthomin, Orthodir and other generalized conjugate gradient schemes. The *generalized minimal residual type algorithms* (GMRES, FGMRES, DQGMRES) are theoretically equivalent and are more robust approaches. One can also maintain the three-term recursion property. This is done by using *biconjugate gradient type algorithms* (BCG, BICGSTAB, DBCG).

The aim of the present article is to discuss the application of the diffuse approximation based collocation method to unsteady fluid flows. In the following sections, the general method of solution is described. Some numerical results obtained by using different preconditioned iterative methods are then given. Two test problems are finally presented. The first is the laminar natural convection in a differentially heated square cavity. The second test case is the oscillatory flow past a circular cylinder.

All numerical simulations have been conducted on a PC computer with 256MB of main memory.

2. The diffuse approximation based collocation method

2.1. *Description of the method*

Let $\Phi: R^n \to R$ be a scalar field whose values Φ_i are known at the points $\mathbf{x}_i$ of a given set of N nodes in the studied domain $D \in R^n$. The diffuse approximation gives estimates of Φ and its derivatives up to the order k at any point $\mathbf{x} \in D$. The Taylor expansion of Φ at $\mathbf{x}$ is estimated using a weighted least squares method which uses only the values of Φ at some points $\mathbf{x}_i$ situated in the vicinity of $\mathbf{x}$.

It can thus be written:

$$\Phi_i^{estimated} = \mathbf{p}(\mathbf{x}_i - \mathbf{x}) \cdot \boldsymbol{\alpha}^{\mathsf{T}}(\mathbf{x}) \tag{1}$$

where $\mathbf{p}(\mathbf{x}_i - \mathbf{x})$ is a line vector of polynomial basis functions and $\boldsymbol{\alpha}(\mathbf{x})$ a vector of coefficients which are determined by minimizing the quantity:

$$I(\boldsymbol{\alpha}) = \sum_{i=1}^{N} \omega(\mathbf{x}, \mathbf{x}_i - \mathbf{x})[\Phi_i - \mathbf{p}(\mathbf{x}_i - \mathbf{x}) \cdot \boldsymbol{\alpha}^{\mathsf{T}}(\mathbf{x})]^2 \tag{2}$$

in which ω is a weight-function of compact support, equal to unity at this point, decreasing when the distance to the node increases and zero outside a given domain of influence (a more precise description of ω will be given next).

Minimization of equation (2) then gives:

$$\mathbf{A}(\mathbf{x})\boldsymbol{\alpha}(\mathbf{x}) = \mathbf{B}(\mathbf{x}) \tag{3}$$

where:

$$\mathbf{A}(\mathbf{x}) = \sum_{i=1}^{N} \omega(\mathbf{x}, \mathbf{x}_i - \mathbf{x})\mathbf{p}^{\mathrm{T}}(\mathbf{x}_i - \mathbf{x})\mathbf{p}(\mathbf{x}_i - \mathbf{x}) \tag{4}$$

$$\mathbf{B}(\mathbf{x}) = \sum_{i=1}^{N} \omega(\mathbf{x}, \mathbf{x}_i - \mathbf{x})\mathbf{p}^{\mathrm{T}}(\mathbf{x}_i - \mathbf{x})\Phi_i \tag{5}$$

In fact $\mathbf{A}(\mathbf{x})$ is the sum of only $n'(\mathbf{x})$ matrix of rank 1, $n'(\mathbf{x})$ being the number of nodes influencing $\mathbf{x}$. By inverting system (3), one obtains the components of $\boldsymbol{\alpha}$ which are the derivatives of Φ at $\mathbf{x}$ in terms of the neighboring nodal values Φ_i. In this work, the Taylor expansion is truncated at order 2. The polynomial vector used is

$$\mathbf{p}(\mathbf{x}_i - \mathbf{x}) = \left\langle 1, (x_i - x), (y_i - y), \frac{(x_i - x)^2}{2}, (x_i - x) \cdot (y_i - y), \frac{(y_i - y)^2}{2} \right\rangle \tag{6}$$

and

$$\langle \alpha_1, \alpha_2, \alpha_3, \alpha_4, \alpha_5, \alpha_6 \rangle = \left\langle \Phi, \frac{\partial \Phi}{\partial x}, \frac{\partial \Phi}{\partial y}, \frac{\partial^2 \Phi}{\partial x^2}, \frac{\partial \Phi^2}{\partial x \partial y}, \frac{\partial^2 \Phi}{\partial y^2} \right\rangle \tag{7}$$

Then, the following system is obtained:

$$\left\{ \begin{array}{c} \varphi \\ \dfrac{\partial \varphi}{\partial x} \\ \dfrac{\partial \varphi}{\partial y} \\ \dfrac{\partial^2 \varphi}{\partial x^2} \\ \dfrac{\partial^2 \varphi}{\partial x \partial y} \\ \dfrac{\partial^2 \varphi}{\partial y^2} \end{array} \right\} = A(X)^{-1} \cdot \left\{ \sum_{i=1}^{n'(X)} \omega(X, X_i - X) \cdot \langle P(X_i - X) \rangle^{\mathrm{T}} \cdot \Phi_i \right\} \tag{8}$$

The square matrix $\mathbf{A}(\mathbf{x})$ is not singular as long as the number $n'(\mathbf{x})$ of the connected nodes at a given point is at least equal to the size of $\boldsymbol{\alpha}$ and are not colinear or cocircular [DEM 84, BRE 02].

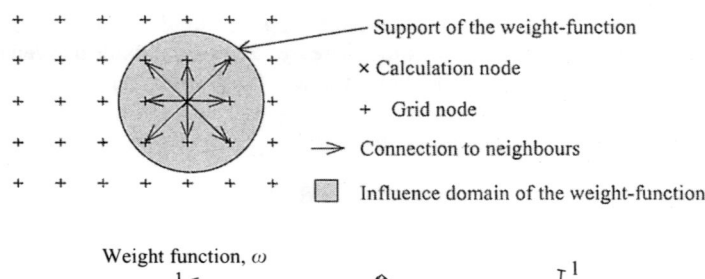

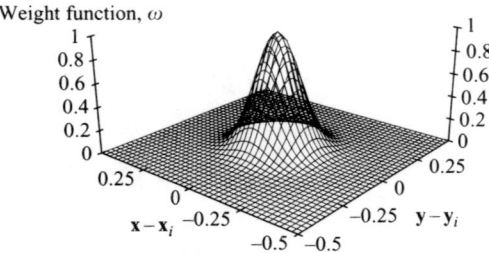

Figure 1 *Neighboring nodes and Gaussian weight function*

In our studies, several weight-functions were tried and it was found that the following Gaussian window (Figure 1):

$$\omega(\mathbf{x}, \mathbf{x}_i - \mathbf{x}) = \exp\left[-3\ln(10) \cdot \left(\frac{|\mathbf{x}_i - \mathbf{x}|}{\sigma}\right)^2\right]$$

$$\omega(\mathbf{x}, \mathbf{x}_i - \mathbf{x}) = 0 \quad \text{if } (\mathbf{x}_i - \mathbf{x})^2 > \sigma^2 \tag{9}$$

behaves rather well. The distance of influence σ is updated at each point in order to use at least 9 neighbors in the approximation.

The previous approximation is then used in a point collocation method to solve partial derivatives equations. At each point of the discretization, the derivatives appearing in the equation to be solved are replaced by their diffuse approximation thus leading to an algebraic system that is solved after the introduction of the Dirichlet boundary conditions. The Neumann boundary conditions on the other hand are replaced by their diffuse approximation and then introduced in the algebraic system as described in [SAD 95, SAD 00, SOP 02].

3. The pressure correction equation

In the primitive variable formulation, the incompressible Navier-Stokes equations (for natural convection problems) can be written as follows [SAD 00]:

$$\frac{\partial \mathbf{v}}{\partial t} + \mathbf{v} \cdot \nabla \mathbf{v} = \frac{\text{Pr}}{\sqrt{\text{Ra}}} \nabla^2 \mathbf{v} - \nabla p + \frac{\mathbf{g}}{g} \text{Pr}\, \theta \tag{10}$$

$$\nabla \cdot \mathbf{v} = 0 \tag{11}$$

$$\frac{\partial \theta}{\partial t} + \mathbf{v}\nabla\theta = \frac{1}{\sqrt{Ra}}\nabla^2\theta. \tag{12}$$

where Pr and Ra are the Prandtl and the Rayleigh numbers respectively, and g is the gravitational acceleration.

Although the pressure gradient term appears in the momentum equation, there is no apparent equation to solve for the pressure. Therefore, special techniques are required. The SIMPLE algorithm [PAT 80] and its various versions and the projection algorithm [COM 82] have been generally used.

These methods are essentially iterative guess-and-correct procedures. They consist of solving the momentum equation by using a guessed at pressure field to obtain an intermediate velocity field. The pressure correction equation, which is obtained by using the continuity equation, is then solved using the intermediate velocity. The process is continued until the convergence test is satisfied.

In this work, we used an equal order projection algorithm, which is described below.

3.1. *Projection algorithm*

The basic methodology of our projection algorithm in the case of bidimensional natural convection can be summarized as follows:

1. Initialization of the fields $(u, v, p, \theta)^i$.
2. Resolution of momentum equations for estimated velocities u^* and v^*.

$$\frac{u^*}{\Delta\tau} - \frac{Pr}{Ra^{1/2}}\left(\frac{\partial^2 u^*}{\partial x^2} + \frac{\partial^2 u^*}{\partial y^2}\right) + u^i\frac{\partial u^*}{\partial x} + v^i\frac{\partial u^*}{\partial y} = \frac{u^i}{\Delta\tau} - \frac{\partial p^i}{\partial x} \tag{13}$$

$$\frac{v^*}{\Delta\tau} - \frac{Pr}{Ra^{1/2}}\left(\frac{\partial^2 v^*}{\partial x^2} + \frac{\partial^2 v^*}{\partial y^2}\right) + u^i\frac{\partial v^*}{\partial x} + v^i\frac{\partial v^*}{\partial y} = \frac{v^i}{\Delta\tau} - \frac{\partial p^i}{\partial y} + Pr\,\theta^i \tag{14}$$

3. Resolution of pressure correction equation

$$\frac{\partial^2 p'}{\partial x^2} + \frac{\partial^2 p'}{\partial y^2} = \frac{1}{\Delta\tau}\left(\frac{\partial u^*}{\partial x} + \frac{\partial v^*}{\partial y}\right) \tag{15}$$

where the boundary condition is:

$$\frac{\partial p'}{\partial \mathbf{n}} = 0$$

on a wall.

4.   Calculation of the correcting component of the velocities.

$$u' = -\frac{\partial p'}{\partial x} \cdot \Delta\tau$$
$$v' = -\frac{\partial p'}{\partial y} \cdot \Delta\tau$$

(16)

5.   Correction of the fields

$$p^{i+1} = p^i + p'$$
$$u^{i+1} = u^* + u'$$
$$v^{i+1} = v^* + v'$$

(17)

6.   Resolution of energy equation.

$$\frac{\theta^{i+1}}{\Delta\tau} + u^{i+1}\frac{\partial\theta^{i+1}}{\partial x} + v^{i+1}\frac{\partial\theta^{i+1}}{\partial y} - \frac{1}{\mathrm{Ra}^{1/2}}\left(\frac{\partial^2\theta^{i+1}}{\partial x^2} + \frac{\partial^2\theta^{i+1}}{\partial y^2}\right) = \frac{\theta^i}{\Delta\tau}$$

(18)

7.   Check for convergence. If the convergence criterion is not respected, we jump to the second step.

3.2. Discretization of the pressure correction equation

During the projection algorithm process, the pressure correction equation (15) is written in its matrix form as follows:

$$\mathbf{M}\mathbf{p}' = \mathbf{b}$$

(19)

where $\mathbf{p}'$ and $\mathbf{b}$ are not to be confused with the previous definitions. The matrix $\mathbf{M}$ is built line by line using equation (8). If $\langle a_l \rangle$ is the lth line of the matrix $\mathbf{A}(\mathbf{x})^{-1}$, and k is the number of the line corresponding to the node $\mathbf{x}_k$ we then have:

$$\mathbf{M}(k, i) = \omega(\mathbf{x}_k, \mathbf{x}_i - \mathbf{x}_k) \cdot (\langle a_4 \rangle + \langle a_6 \rangle) \cdot \mathbf{p}(\mathbf{x}_i - \mathbf{x}_k)$$

(20)

and

$$\mathbf{b}(k) = \frac{1}{\Delta\tau}\left[\sum_{i=1}^{n'(\mathbf{x}_k)} \omega(\mathbf{x}_k, \mathbf{x}_i - \mathbf{x}_k) \cdot (\langle a_2 \rangle \cdot u_i^* + \langle a_3 \rangle \cdot v_i^*) \cdot \mathbf{p}(\mathbf{x}_i - \mathbf{x}_k)^{\mathrm{T}}\right]$$

(21)

where u^* and v^* are the estimated values of the velocity.

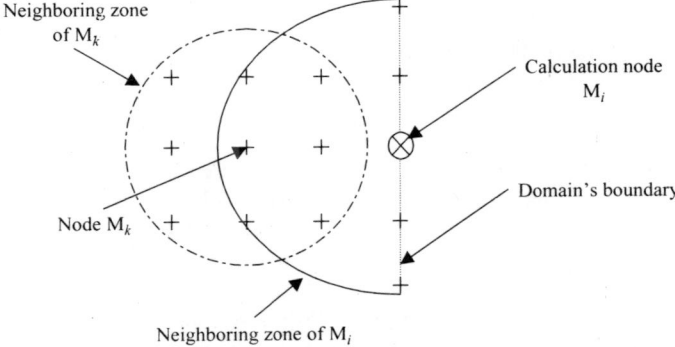

Neighboring zone
of M_k

Calculation node
M_i

Domain's boundary

Node M_k

Neighboring zone of M_i

Figure 2 *Boundary nodes and their neighbours*

The Gaussian weight function used in this work is shown in Figure 1. Let us consider now the situation depicted in Figure 2 where the node M_i is localized on a boundary, and where M_k is a neighboring node of M_i. During the discretization process at M_k the node M_i is not involved whereas the implementation of the Neumann type boundary condition at the node M_i involves the node M_k. This leads to an asymmetrical matrix, even if the non-symmetric element number is very low compared to the total number of elements. This leads to a slow convergence of the iterative algorithm.

4. Performance evaluation of iterative methods

One of the important properties of CG-like methods is the so-called super linear convergence behaviour. The convergence rate improves as the iteration proceeds. In many cases, the initial convergence can be very irregular and slow. Therefore, the high asymptotic convergence rate may not be so desirable if the early stage convergence is slow or unstable. This motivates this study to investigate the early stage convergence behaviour of various CG-like methods when applied to find the numerical solution of the pressure correction system that arises in our projection algorithm.

The following methods were tested with the pressure correction equation matrix obtained in the differentially heated square cavity problem:

- Bi-Conjugate Gradient (BCG)
- Bi-Conjugate Gradient with partial Pivoting (DBCG)
- Conjugate Gradient for Normal Residual Equation (CGNR)
- Bi-Conjugate Gradient Stabilized (BCGSTAB)
- Transpose-Free Quasi-Minimum Residual method (TFQMR)
- Generalized Minimum Residual (GMRES)
- Flexible version of GMRES (FGMRES)

– Direct Quasi-GMRES (DQGMRES)
– Full Orthogonal Method (FOM)

and the results are given in the following section.

A description of these methods can be found elsewhere [SAA 96]. It is well known that the use of a preconditioner improves considerably the convergence process. One of the most used technique is incomplete LU factorization with different fill levels ILU(k). We can also mention modified incomplete LU factorization, MILU(k). In this work, we have chosen the ILUT(k) preconditioner which was implemented as suggested by Saad [SAA 96]. Two parameters corresponding to the number of elements kept on each line of the matrix (excepting the diagonal values), and the value under which elements are ignored, are respectively set to (lfil = 15) and (droptol = 10^{-4}).

4.1. Problem description

The test problem originates from the simulation of laminar natural convection in a differentially heated square cavity (Figure 3a). In this problem, the fluid reaches a steady state flow for a wide range of Rayleigh number (up to 10^8). At a critical value around Ra = $1.8.10^8$, the system undergoes a first bifurcation to a pseudo periodic solution. We have thus chosen to test the different algorithms at a Rayleigh number of 10^8 whose solution is depicted in Figure 3b.

The domain is discretized with different irregular grids (from 41×41 up to 201×201). In the x, y plane, the non uniform grids obeyed the law:

$$\begin{pmatrix} x \\ y \end{pmatrix} = \frac{1}{2} \left\{ 1 + \frac{\tanh\left(2\left(\dfrac{x_r}{y_r}\right) - 1\right)}{\tanh(1)} \right\} \tag{22}$$

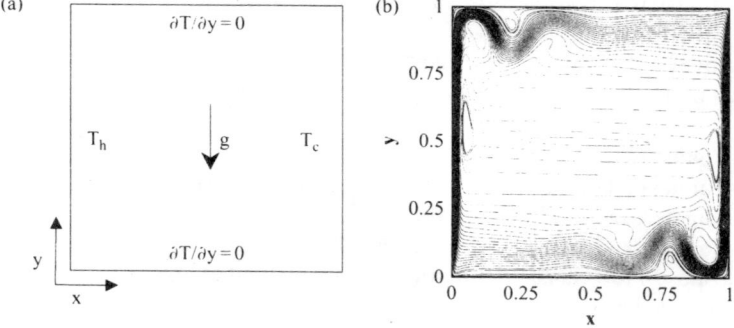

Figure 3 (a) Differentially heated square cavity (b) streamlines for Ra = 10^8

where the variables x_r, y_r change within $[0, 1]$ and are uniformly distributed. The choice of a finer grid near the walls is motivated by the need to improve the solution in the boundary layers. The time step is fixed to 2.10^{-2} for all the simulations.

4.2. *Results*

In this section, we present some results obtained by using the pressure matrix equation at the first iteration of the first time step. The maximum number of iterations has been fixed at 1000 for all the grids used except for the 201×201 grid for which a number of 3000 has been used. If the maximum number of iterations is reached, the iterative solver stops and returns the appropriate warning. The vector **b** used in equation (19) is set in order to obtain a unity vector solution (so the error can be easily estimated).

The convergence criterion is defined as follow:

$$\|\mathbf{A}\mathbf{x}_i - \mathbf{b}\| \leq rtol \cdot \|\mathbf{A}\mathbf{x}_1 - \mathbf{b}\| + atol \tag{23}$$

where *rtol*, *atol* are two parameters that are fixed for each case, and $\mathbf{x}_1$, $\mathbf{x}_i$ are solutions approached at the first and at the ith iteration. Five different cases are considered during this study (Table 1).

The iteration number necessary to reach the convergence criterion and the residual norm ($\|\mathbf{A}\mathbf{x} - \mathbf{b}\|$) are given in Table 2 for each case.

It can be seen that the CGNR algorithm is very slow whenever it converges, while BCG and DBCG algorithms have similar comportment for all the treated cases. Concerning GMRES type methods and FOM, one can see that they are faster for low order systems (41×41 grid). For the 81×81 and the 201×201 grids, BCGSTAB and TFQMR appear to be more efficient (BCGSTAB being the fastest).

Although it is not among the aims of the present article to study the problem of transition to non steady flow for the cavity problem, it is useful to point out that an unsteady solution has been found at Ra $= 2.10^8$ using a 201×201 mesh. We found a fundamental frequency of 0.0518, which is in close agreement with the findings

Table 1 *Description of test cases*

	Ra	Grid	*rtol*	*atol*
Case 1	10^8	81×81	10^{-10}	0
Case 2	10^8	41×41	10^{-10}	0
Case 3	10^8	41×41	0	10^{-6}
Case 4	10^8	201×201	10^{-10}	0
Case 5	10^8	201×201	0	10^{-6}

Table 2 *Convergence results*

Method	Case 1		Case 2		Case 3		Case 4		Case 5	
	Iteration number	Residual norm	Iteration number	Residual norm	Iteration number	Residual norm	Iteration number	Residual norm	Iteration number	Residual norm
Bcg	68	$1.61.10^{-7}$	34	$1.32.10^{-7}$	32	$1.29.10^{-7}$	178	$1.16.10^{-5}$	186	$4.85.10^{-7}$
Dbcg	69	$1.61.10^{-7}$	35	$1.32.10^{-7}$	33	$1.19.10^{-7}$	201	$1.19.10^{-5}$	231	$2.76.10^{-7}$
–	–	–	810	$4.6.10^{-7}$	–	–	–	–	–	–
47	–	$1.29.10^{-6}$	22	$3.4.10^{-7}$	–	–	–	–	–	–
Cgnr	–	–	–	–	799	$8.15.10^{-7}$	–	–	–	–
Bcgstab	–	–	–	–	21	$3.4.10^{-7}$	139	$5.1.10^{-6}$	143	$6.37.10^{-7}$
Tfqmr	57	$4.42.10^{-7}$	26	$1.18.10^{-7}$	25	$1.25.10^{-7}$	205	$1.11.10^{-6}$	207	$5.32.10^{-7}$
Gmres	107	$1.81.10^{-6}$	18	$5.62.10^{-8}$	16	$7.01.10^{-7}$	1056	$1.46.10^{-5}$	1245	$9.13.10^{-7}$
Fgmres	107	$1.81.10^{-6}$	18	$5.62.10^{-8}$	" "	" "	1000	$1.46.10^{-5}$	1207	$9.2.10^{-7}$
Dqgmres	74	$1.7.10^{-6}$	18	$5.62.10^{-8}$	" "	" "	876	$1.42.10^{-5}$	929	$8.93.10^{-7}$
Fom	92	$1.35.10^{-6}$	18	$5.62.10^{-8}$	" "	$7.04.10^{-7}$	1018	$1.31.10^{-5}$	1255	$9.28.10^{-7}$

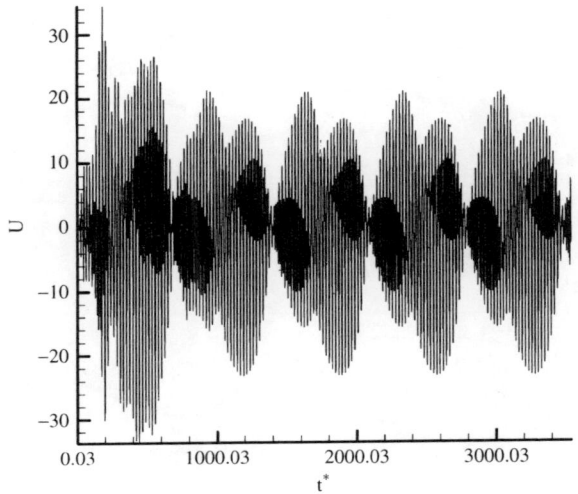

Figure 4 *Time evolution of the horizontal velocity at point (x = 0.5, y = 0.5)*

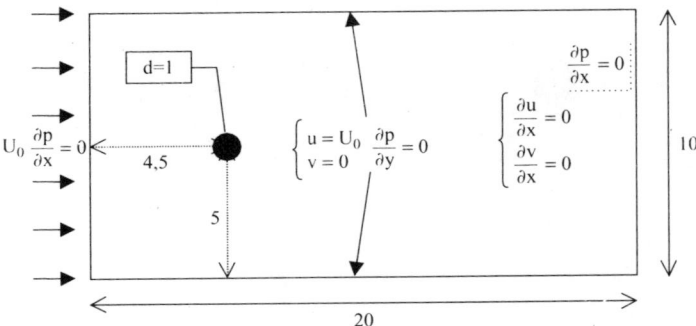

Figure 5 *Description of the cylinder problem*

($f = 0.0522$) of Janssen *et al.* [JAN 93]. The time evolution of the velocity at $(x = 0.5; y = 0.5)$ is finally depicted in Figure 4.

5. Flow around a circular cylinder

The fluid flow past a circular cylinder is the second case considered. The problem description and boundary conditions are shown in Figure 5.

The flow has a steady state solution composed of two contrarotative vortexes for Reynolds numbers up to Re $\approx$ 35. Above this critical Reynolds number, the two

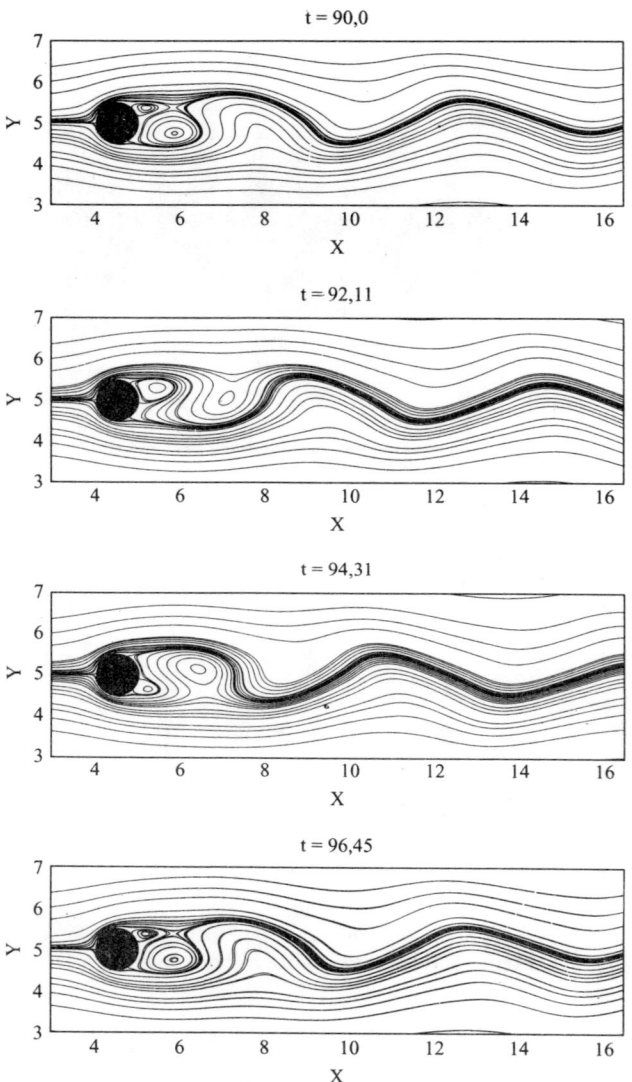

Figure 6 *Streamlines over a period for* Re $= 65$

cells start to oscillate and lengthen successively, making a fluid detachment at a frequency f related to the Strouhal number (St $= f \cdot D/U_\infty$).

The flow has been simulated for a Reynolds number Re $= 65$ with a 30 000 nodes irregular grid and an adimensional time step $\Delta\tau = 0.02$. The calculated Strouhal number St $= 0.155$ compares very well with the results of Saiki *et al.* [SAI 96] who found St = 0.152 with a 64 000 nodes grid and a virtual boundary method. The streamlines over a period are shown in Figure 6.

6. Conclusion

A diffuse approximation method for solving unsteady incompressible fluid flow problems has been presented. It was demonstrated that preconditioned BICGSTAB is a suitable method for the solution of the pressure correction equation. As shown by the comparison with existing numerical solutions, results are very accurate in both space and time. We have not discussed here the problem of essential boundary conditions which remains an open question still. Further work is still needed in that direction.

7. References

[ALU 01] Aluru N. R., Li G., Finite Cloud Method: A True Meshless Technique Based on Fixed Reproducing Kernel, *International Journal for Numerical Methods in Engineering*, Vol. 50, 2001, pp. 2373–2410.

[BEL 94] Belytschko T., Lu Y.Y., Gu L., Element-free Galerkin Methods, *International Journal for Numerical Methods in Engineering*, Vol. 37, 1994, pp. 229–256.

[BRE 02] Breitkopf P., Rassineux A. and Villon P., An Introduction to the Moving Least Squares Meshfree Methods, *Revue Européenne des Éléments Finis*, No 7–8, 2002.

[CHA 01] Chati M. K., Paulino G. H., Mukherjee S., The Meshless Standard and Hypersingular Boundary Node Methods-applications to Error Estimation and Adaptivity in the Three-dimensional Problems, *International Journal for Numerical Methods in Engineering*, Vol. 50, 2001, pp. 2233–2269.

[COM 82] Comini G. and Del Giudice S., Finite Element Solution of the Incompressible Navier-Stokes Equations, *Num. Heat Transfer*, Vol. 5, 1982, pp. 463–478.

[COU 98] Couturier S., Sadat H., Résolution des Equations de Navier-Stokes dans la Formulation en Variables Primitives par Approximation Diffuse , *C. R. Acad. Sciences*, t. 326, série IIb, 1998, pp. 117–119.

[DEM 84] Demkowicz L., Karafiat A. and Liszka T., On some Convergence Results for FDM with Irregular Mesh , *Comp. Methods Appl. Mech. Engg.*, Vol. 42, 1984, pp. 343–355.

[JAN 93] Janssen R. J. A., Henkes R. A. W. M., Accuracy of Finite-volume Discretizations for Bifurcating Natural-convection Flow in a Square Cavity, *Numerical Heat Transfer*, Part B, Vol. 24, 1993, pp. 191–207.

[LIU 01] Liu G. R., Gu Y. T., Local Point Interpolation Method for Stress Analysis of Two-dimensional Solids, *Structural Engineering and Mechanics*, Vol. 11, 2001, pp. 221–236.

[OHS 01] Ohs R. R., Aluru N. R., Meshless Analysis of Piezoelectric Devices , *Computational Mechanics*, Vol. 27, 2001, pp. 23–36.

[PAT 80] Patankar S. V., *Numerical Heat Transfer and Fluid Flow*, Hemisphere, Washington, DC 1980.

[PRA 98] Prax C., Salagnac P. and Sadat H., Diffuse Approximation Method and Control Volume-based Finite Element Methods : a Comparative Study, *Num. Heat Transfer*, Part B, Vol. 34, 1998, pp. 303–321.

[SAA 96] Saad Y., *Iterative Methods for Sparse Linear System*, PWS, Boston, MA.

[SAD 95] Sadat H., Prax C., Résolution des Problèmes Mécanique des Fluides et de Thermique par Approximation Diffuse, *Congrès de la Société Française des Thermiciens*, Poitiers, mai 1995.

[SAD 96] Sadat S., Prax C., Salagnac P., Diffuse Approximation Method for Solving Natural Convection in Porous Media, *Transport in Porous Media*, Vol. 22, No. 2, 1996, pp. 215–223.

[SAD 00] Sadat H. and Couturier S., Performance and Accuracy of a Meshless Method for Laminar natural Convection, *Numer. Heat Transfer, Part B Fundamentals*, Vol. 37, 2000, pp. 455–467.

[SAI 96] Saiki E. M., Biringen S., Numerical Simulation of a Cylinder in Uniform Flow: Application of a Virtual Boundary Method, *J. Comput. Phys.*, Vol. 123, 1996, pp. 450–465.

[SOP 02] Sophy T., Sadat H., Prax C., A Meshless Formulation for Three-dimensional Laminar Natural Convection, *Num. Heat Transfer, Part B Fundamentals*, Vol. 41, 2002, pp. 433–445.

[ZHA 00] Zhang X., Song K. Z., Lu M. W., Liu X., Meshless Methods Based on Collocation with Radial Basis Functions, *Computational Mechanics*, Vol. 26, 2000, pp. 333–343.

Index